T0187080

Cambridge Lower Secondary

Maths

STAGE 7: STUDENT'S BOOK

Alastair Duncombe, Rob Ellis, Amanda George, Claire Powis, Brian Speed

Series Editor: Alastair Duncombe

Collins

William Collins' dream of knowledge for all began with the publication of his first book in 1819.

A self-educated mill worker, he not only enriched millions of lives, but also founded a flourishing publishing house. Today, staying true to this spirit, Collins books are packed with inspiration, innovation and practical expertise. They place you at the centre of a world of possibility and give you exactly what you need to explore it.

Collins. Freedom to teach.

Published by Collins
An imprint of HarperCollins*Publishers*
The News Building
1 London Bridge Street
London
SE1 9GF

HarperCollins*Publishers*
Macken House, 39/40 Mayor Street Upper,
Dublin 1, D01 C9W8, Ireland

Browse the complete Collins catalogue at
www.collins.co.uk

This book is produced from independently certified FSC™ paper to ensure responsible forest management.

For more information visit:
www.harpercollins.co.uk/green

British Library Cataloguing in Publication Data
A catalogue record for this publication is available from the British Library.

Authors: Alastair Duncombe, Rob Ellis, Amanda George, Mark Heslop, Claire Powis, Peter Ransom, Brian Speed
Series editor: Alastair Duncombe

Publisher: Elaine Higgleton
In-house project editors: Jennifer Hall and Caroline Green
Project manager: Wendy Alderton
Development editors: Anna Cox, Rachel Hamar and Phil Gallagher
Copyeditor: Alison Bewsher
Proofreader: Eric Pradel
Answer checker: Eric Pradel and Jouve India Private Limited
Cover designer: Ken Vail Graphic Design and Gordon MacGlip
Cover illustrator: Ann Paganuzzi
Typesetter: Jouve India Private Limited
Production controller: Lyndsey Rogers
Printed and bound in India by Replika Press Pvt. Ltd.

Acknowledgements

The publishers gratefully acknowledge the permission granted to reproduce the copyright material in this book. Every effort has been made to trace copyright holders and to obtain their permission for the use of copyright material. The publishers will gladly receive any information enabling them to rectify any error or omission at the first opportunity.

p. 91 The World Bank: People using safely managed drinking water services (% of population): WHO/UNICEF Joint Monitoring Programme (JMP) for Water Supply, Sanitation and Hygiene (washdata.org). Licence : CC BY-4.0; p.114 daizuoxin/Shutterstock; p. 135 UNdata http://data.un.org/_Docs/SYB/PDFs/SYB61_253_Population%20Growth%20Rates%20in%20Urban%20areas%20and%20Capital%20cities.pdf; p. 228 Clean Energy Wire, made available under a Creative Commons Attribution 4.0 International Licence https://creativecommons.org/licenses/by/4.0/; p. 231 Global EV Outlook 2018, https://www.iea.org/reports/global-ev-outlook-201; p.232 BP Statistical Review of World Energy 2020; p. 232 United Nations, Department of Economic and Social Affairs, Population Division (2019). World Population Ageing 2019: Highlights (ST/ESA/SER.A/430). Copyright © 2019 by United Nations, made available under a Creative Commons license (CC BY 3.0 IGO) http://creativecommons.org/licenses/by/3.0/igo/; p233 UN Data; p. 233 Worldometers.info https://www.worldometers.info/water; p. 244 National Records of Scotland © Crown copyright and database right 2018; p.248 Ivan Baranov/Shutterstock; p.280 Lorna Roberts/Shutterstock; p. 284 robuart/Shutterstock; p. 302 Lisa-S/Shutterstock; p. 311 Teerasak Ladnongkhun/Shutterstock

Cambridge International copyright material in this publication is reproduced under licence and remains the intellectual property of Cambridge Assessment International Education.

Third-party websites and resources referred to in this publication have not been endorsed by Cambridge Assessment International Education.

With thanks to the following contributors from the first edition: Michele Conway, Sarah Sharratt and Deborah McCarthy.

With thanks to the following teachers and schools for reviewing materials in development: Samitava Mukherjee and Debjani Sen, Calcutta International School; Hawar International School; Adrienne Leisztinger, International School of Budapest; Sujatha Raghavan, Manthan International School; Mahesh Punjabi, Podar International School; Taman Rama Intercultural School; Utpal Sanghvi International School.

Introduction

The *Collins Lower Secondary Maths Stage 7 Student's Book* covers the Cambridge Lower Secondary Mathematics curriculum framework (0862). The content has been covered in 27 chapters. The series is designed to illustrate concepts and provide practice questions at a range of difficulties to allow you to build confidence on a topic.

The authors have included plenty of worked examples in every chapter. These worked examples will lead you, step-by-step, through the new concepts. They include clear and detailed explanations. Where possible, links have been made between topics, encouraging you to build on what you know already, and to practise mathematical concepts in a different context.

Each chapter within the book contains activities and questions to help you to develop your skills with *thinking and working mathematically*. You will practice the skills of *specialising* and *generalising*, *conjecturing* and *convincing*, *characterising* and *classifying*, and *critiquing* and *improving*. You can find definitions of each of these characteristics on the next page. The activities and questions will help you to understand each topic. They will also develop your skills at spotting patterns and solving mathematical problems. These activities and questions are indicated by a star:

Every chapter has these helpful features:
- 'Starting point': to remind you of what you know already

- 'This will also be helpful when ...': to let you know where you will use the mathematics in the future

- 'Getting started': to get you interested in the new topic through an activity

- 'Key terms' boxes: to identify new mathematical words you need to know in that chapter, and provide a definition

- Clear topic headings: so that you can see what you are going to be learning in each section of the chapter

- Worked examples: to show you how to answer questions with both formal and informal (diagrammatic) explanations provided

- 'Tip' boxes: to give you guidance on the possible methods and common errors

- Exercises: to give you practice at answering questions on each topic. The questions at the end of each exercise will be harder, in order to stretch you

- 'Thinking and working mathematically' questions and activities (marked as ▼): to help you develop your mathematical thinking. The activities will often be more open-ended in nature

- 'Think about' boxes: to suggest ideas that you might want to consider

- 'Discuss' boxes: to encourage you to talk about mathematical ideas with a partner or in class

- 'Did you know?' boxes: to explain where mathematical ideas came from and how they are applied in real life.

- 'Consolidation exercise': to give you further practice on all the topics introduced in the chapter.

- 'End of chapter reflection': to help you think about how well you have understood the ideas in the chapter, so that you can monitor your own progress.

We hope that you find this approach enjoyable and engaging as you progress through your mathematical journey.

The Thinking and Working Mathematically Star

Critiquing

Comparing and evaluating mathematical ideas, representations or solutions to identify advantages and disadvantages.
For example:

◄ Which is the best way to …?

◄ Write down the advantages and disadvantages of …

Specialising

Choosing an example and checking to see if it satisfies or does not satisfy specific mathematical criteria.
For example:

◄ Find an example of …

◄ Find … if … and …

Conjecturing

Forming mathematical questions or ideas.
For example:

◄ What would happen if …?

◄ How would you …?

Specialising and Generalising

Conjecturing and Convincing

Critiquing and Improving

Generalising

Recognising an underlying pattern by identifying many examples that satisfy the same mathematical criteria.
For example:

◄ Find a rule that connects … and …

◄ What can you conclude from …?

Convincing

Presenting evidence to justify or challenge a mathematical idea or solution.
For example:

◄ Prove that …

◄ Explain why …

◄ Show that …

Characterising and Classifying

Characterising

Identifying and describing the mathematical properties of an object.
For example:

◄ What do … have in common?

◄ Describe the properties of …

Improving

Refining mathematical ideas or representations to develop a more effective approach or solution.
For example:

◄ Find a better way to …

◄ Describe a more efficient way to …

Classifying

Organising objects into groups according to their mathematical properties.
For example:

◄ Match …

◄ Sort …

◄ Put a ring around all the … which …

The Thinking And Working Mathematically Star, © Cambridge International, 2018

Contents

1 Factors

You will learn how to:
- Use knowledge of tests of divisibility to find factors of numbers greater than 100.
- Understand lowest common multiple and highest common factor (numbers less than 100).

Starting point

Do you remember…

- how to test a number less than 100 for divisibility by 2, 3, 4, 5, 6, 8, 9, 10, 25 and 50?

  For example, find out whether 96 is divisible by 2, 3, 4, 5, 6, 8, 9, 10, 25 or 50.
- how to find common multiples of two numbers?

  For example, find three common multiples of 3 and 4.
- how to find common factors of two numbers?

  For example, find all the common factors of 12 and 18.

This will also be helpful when…

- you find the prime factors of whole numbers
- you find the lowest common multiple and highest common factor of numbers greater than 100.

1.0 Getting started

Every recently published book has a unique 13-digit number, called an ISBN (International Standard Book Number). You can usually find it on the back cover or on one of the first pages.

You can ignore the lines between the digits. The first 12 digits are a code which uniquely identifies the book. The last digit is a 'check digit'. It is not chosen – it is calculated:

1 Multiply the first digit by 1, the second digit by 3, the third digit by 1, the fourth digit by 3. Continue this pattern until you multiply the eleventh digit by 1 and the twelfth digit by 3.

2 Find the sum of all of these calculations.

3 The check digit is related to this sum – see below.

Below are some more examples of possible ISBNs. Can you find how the check digit is worked out?

978-0-33-444444-2

978-1-11-111111-0

978-2-18-160034-7

The check digit is used to avoid mistakes when reading the ISBN, for example when using a barcode reader or ordering books online.

The check digit is the is the smallest number that needs to be added to the sum from step 2, to make a number that is divisible by 10.

- Find the ISBN of this book. Show that its check digit is correct.
- Here are some examples of possible ISBNs, but with the check digit missing. Find the missing check digit.
  978-0-33-444444-?
  978-1-11-111111-?
  978-2-18-160034-?

Did you know?

The check digit is used to avoid mistakes when reading an ISBN, for example when using a barcode reader or when ordering books online.

- Choose three ISBNs, choosing the first twelve digits and then calculating the check digit. Change one of the check digits to an incorrect value. Ask another student to work out which ISBN is incorrect.

- Can you create a spreadsheet that works out the check digit from the first twelve digits of an ISBN?

1.1 Divisibility tests

Key terms

A number is **divisible** by another number if it can be divided exactly by that number without leaving a remainder. For example, 12 is divisible by 6 because $12 \div 6 = 2$ with no remainder.

A **divisibility test** is a quick method for checking whether one number is divisible by another number.

A number is divisible by:		Examples
2	if the number is even. Even numbers end in 0, 2, 4, 6 or 8.	18, 770, 5806
3	if the sum of the digits is divisible by 3.	6225 ($6 + 2 + 2 + 5 = 15$, which is divisible by 3)
4	if the number made by the last two digits is divisible by 4.	9584 ($84 \div 4 = 21$ with no remainder)
5	if the last digit is 5 or 0.	430, 8475
6	if the number is divisible by both 2 and 3.	3492 (ends in 2, which is even, and $3 + 4 + 9 + 2 = 18$, which is divisible by 3)
7	(There is no simple divisibility test for 7.)	
8	if the number made by the last three digits is divisible by 8.	1224 ($224 \div 8 = 28$ with no remainder)
9	if the sum of the digits is divisible by 9.	9216 ($9 + 2 + 1 + 6 = 18$, which is divisible by 9)
10	if the last digit is 0.	900, 6710
25	if the last two digits are 00, 25, 50 or 75.	575, 4825
50	if the last two digits are 00 or 50.	600, 7750
100	if the last two digits are 00.	900, 4100

Think about

What is zero divisible by?

Is 9000 divisible by 8?
Explain your answer.

Worked example 1

Test the number 3426 for divisibility by 3, 4, 5, 6, 8 and 9.

3426 is divisible by 3	3 + 4 + 2 + 6 = 15 15 is divisible by 3	Find the sum of the digits. If this is divisible by 3, then the original number is also divisible by 3.
3426 is not divisible by 4	26 is not divisible by 4	Look at the final 2 digits and see if they are divisible by 4.
3426 is not divisible by 5	3426 does not end in 0 or 5	All multiples of 5 end in either 5 or 0.
3426 is divisible by 6	3426 is an even number so is divisible by 2. It is also divisible by 3.	Check to see if the number is divisible by both 2 and 3.
3426 is not divisible by 8	426 divided by 8 is 53 remainder 2	Look at the final 3 digits and see if they are divisible by 8.
3426 is not divisible by 9	3 + 4 + 2 + 6 = 15, which is not divisible by 9	Find the sum of the digits. If this is divisible by 9, then the original number is also divisible by 9.

Exercise 1

1 Look at these numbers.

   127   144   168   219   744

  a) Which numbers are divisible by 2?

  b) Which numbers are divisible by 3?

  c) Which numbers are divisible by 6?

2 Look at these numbers.

   135   360   900   1248   6700

  a) Which numbers are divisible by 5?

  b) Which numbers are divisible by 10?

  c) Which numbers are divisible by 100?

3 Look at these numbers.

   542   639   846   1413   1732

  a) Which numbers are divisible by 9?

  b) Which numbers are divisible by 4?

4 Which of these numbers is 3600 divisible by?

  2, 3, 4, 5, 6, 8, 9, 10, 100

5 Write down which of these numbers 6075 is divisible by:

4, 5, 8, 9, 10, 25, 50

6 Write down which of these numbers 12 848 is divisible by:

3, 4, 5, 6, 8, 9, 10

7 Here is a test for divisibility by 7:

Double the last digit and subtract from the number formed from the other digits.
If the answer is divisible by 7, then so was the original number.

For example: 376
$37 - (2 \times 6) = 37 - 12 = 25$.
25 is not divisible by 7, so 376 is not divisible by 7.

Use this test to find which of these numbers is divisible by 7:

a) 182   **b)** 223   **c)** 406   **d)** 426   **e)** 619

8 Find a number that is divisible by 1, 2, 3, 4, 5, 6, 7 and 8. How did you do this?

Is this the smallest possible number that is divisible by 1, 2, 3, 4, 5, 6, 7 and 8?
Explain your answer.

9 Using the idea of divisibility, classify the numbers below into two groups. Explain your choices.
(There is more than one way to do this.)

2088   2232   2328   2388   3144   3156   3636   3924

Thinking and working mathematically activity

If a number is divisible by 2 and 3, it is also divisible by their product, 6. Choose some numbers that are divisible by 2 and 3. Show that all of these numbers are divisible by 6. Try to explain this.

If a number is divisible by 2 and 4, it is not necessarily divisible by their product, 8. Choose some numbers that are divisible by 2 and 4. Show that some of these numbers are divisible by 8 and some are not. Try to explain this.

Investigate other pairs of factors. For some factor pairs, all numbers divisible by both factors are also divisible by their product. What do these factor pairs have in common?

1.2 Lowest common multiple and highest common factor

Key terms

The lowest common multiple (LCM) of two or more whole numbers is the lowest multiple shared by the numbers. For example, 4 and 5 have common multiples of 20, 40, 60, and so on. The lowest common multiple of 4 and 5 is 20.

The highest common factor (HCF) of two or more whole numbers is the highest factor shared by the numbers. For example, 16 and 24 have common factors 1, 2, 4 and 8. The highest common factor of 16 and 24 is 8.

A **product** is the result of multiplying two or more numbers. For example, the product of 2 and 3 is 6.

Did you know?

A factor of a number is sometimes called a divisor.

The highest common factor is sometimes called the greatest common divisor (GCD).

Worked example 2

a) Find the lowest common multiple of 6 and 10.

b) Find the highest common factor of 18 and 27.

a) 6, 12, 18, 24, **30**, 36 … 10, 20, **30**, 40, 50 … Lowest common multiple = 30	Write a list of multiples of 6, and a list of multiples of 10. Continue writing multiples until you find the smallest number that is in both lists. This is the lowest common multiple.	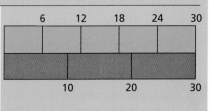
b) 18: 1, 2, 3, 6, **9**, 18 27: 1, 3, **9**, 27 Highest common factor = 9	Write all the factors of 18 and all the factors of 27. The largest number that appears in both lists is the highest common factor.	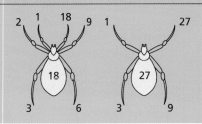

Exercise 2

Think about

How can you use a lowest common multiple to help you add the fractions $\frac{1}{4}$ and $\frac{2}{3}$?

1 a) Which of these numbers are common multiples of 6 and 8?

      6   8   12   24   36   48

  b) Which of these numbers are common multiples of 10 and 15?

      5   20   30   40   45   60

  c) Which of these numbers are common multiples of 12 and 24?

      6   24   36   60   72   84

2 Find the lowest common multiple of:

a) 5 and 7

b) 2 and 4

c) 3 and 11

d) 6 and 15

e) 10 and 12

f) 14 and 21

Thinking and working mathematically activity

Sometimes the lowest common multiple of two numbers is the product of the two numbers. For example, the LCM of 5 and 9 is $5 \times 9 = 45$. Make a list of pairs of numbers whose LCM is their product.

Sometimes the lowest common multiple is not the product of the two numbers. For example, the LCM of 6 and 9 is not $6 \times 9 = 54$; it is 18. Make a list of pairs of numbers whose LCM is not their product.

Look at your two lists. What do the pairs of numbers in each list have in common? Try to explain this.

---

3 Find the lowest common multiple of:

a) 18 and 36

b) 20 and 25

c) 16 and 40

d) 40 and 60

4 Find the lowest common multiple of:

a) 2, 3 and 5

b) 3, 6 and 9

c) 4, 5 and 6

d) 5, 15 and 20

5 a) Which of these numbers are common factors of 15 and 45?

1   3   5   7   9   15

b) Which of these numbers are common factors of 20 and 100?

4   5   10   20   50   100

c) Which of these numbers are common factors of 24 and 36?

2   3   4   6   8   12

6 Praneeth has used the Venn diagram shown to find the common factors of two numbers. He has made some errors. Find the errors and rewrite his Venn diagram.

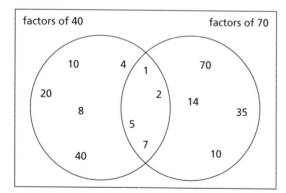

7 a) Draw a Venn diagram, similar to the one in question 6, to show the factors of 16 and the factors of 20. Write common factors in the overlapping section.

b) Find the highest common factor of 16 and 20.

8 Find the highest common factor of:

a) 4 and 6

b) 10 and 15

c) 13 and 21

d) 14 and 24

e) 14 and 28

f) 27 and 36

---

9 Find the highest common factor of:

   a) 24 and 80   **b)** 30 and 75

   c) 32 and 100   **d)** 40 and 88

10 Find the highest common factor of:

   a) 9, 15 and 30   **b)** 24, 32 and 80   **c)** 33, 66 and 77

11 a) Find two pairs of numbers such that each pair has highest common factor 3 and lowest common multiple 18. Describe your method.

   b) Compare your method with other students' methods. Explain which method you think is most efficient for solving this problem.

   c) Write your own question similar to part **a)**, and show how to solve it.

> **Think about**
>
> To find the highest common factor of three numbers, such as 8, 16 and 96, do you always need to find all of the factors of all three numbers?

Consolidation exercise

1 The students in a classroom can be divided into five equal groups or into six equal groups. Work out the smallest possible number of students in the classroom.

2 Tabitha goes swimming every four days and Saliha goes swimming every six days. They both go swimming today. Find the number of days until the next time they swim on the same day.

> **Discuss**
>
> 'The lowest common multiple of two numbers is always a multiple of the highest common factor.' Is this statement true or false?

3 x is a whole number. The highest common factor of x and 8 is 4. The lowest common multiple of x and 5 is 60. Find x.

4 A wall measures 60 centimetres by 100 centimetres. It is completely covered by identical square tiles. Find the smallest possible number of tiles on the wall.

5 Liam makes 32 chocolate biscuits and 40 ginger biscuits. He puts the biscuits into boxes. Each box contains the same number of chocolate biscuits and the same number of ginger biscuits.

   a) Find the largest possible number of boxes.

   b) Work out the number of chocolate biscuits and the number of ginger biscuits in each box.

6 Use divisibility tests to answer the questions below. Show your methods.

   a) Is 5172 divisible by 6?   **b)** Is 6410 divisible by 15?

7 Use each of the numbers from 0 to 9 exactly once in the empty boxes to make each of the following statements true.

43 ☐ is divisible by 5 and ☐

6 ☐ 2 is divisible by 4

117 ☐ 0 is divisible by ☐ 00 and ☐

A number that is divisible by 3 and ☐ is also divisible by ☐

☐ 2 is divisible by ☐

Think about

Is there more than one answer to this problem?

End of chapter reflection

You should know that...	You should be able to...	Such as...
The lowest common multiple (LCM) of two or more numbers is the lowest multiple that the numbers have in common.	Find the LCM of two or three numbers that are less than 100.	Find the LCM of 9 and 12. Find the LCM of 3, 6 and 14.
The highest common factor (HCF) of two or more numbers is the highest factor that the numbers have in common.	Find the HCF of two or three numbers that are less than 100.	Find the HCF of 16 and 40. Find the HCF of 12, 20 and 52.
There are divisibility tests for the numbers 2, 3, 4, 5, 6, 8, 9, 10, 25, 50 and 100.	Test whether a number over 100 is divisible by 2, 3, 4, 5, 6, 8, 9, 10, 25, 50 or 100.	Is 31527 divisible by 9?

2 2D and 3D shapes

You will learn how to:
- Understand that if two 2D shapes are congruent, corresponding sides and angles are equal.
- Know the parts of a circle:
  - centre
  - radius
  - diameter
  - circumference
  - chord
  - tangent.
- Identify and describe the combination of properties that determine a specific 3D shape.

Starting point

Do you remember…

- what a side is and what vertices are?

  For example, how many sides and vertices does a quadrilateral have?

- the angles of a triangle add up to 180°?

  For example, find the size of the angle marked x in this triangle.

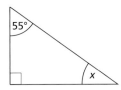

- how to draw a circle with a pair of compasses?

  For example, draw a circle with a diameter of 5 cm.

- what parallel lines are?

  For example, find a pair of parallel lines in this diagram.

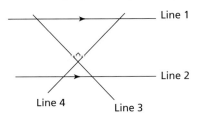

- the names of common 3D shapes?

  For example, write down the name of this shape.

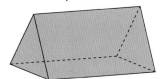

This will also be helpful when…

- you transform shapes by reflection, translation and rotation.

2.0 Getting started

Here is a tangram puzzle.

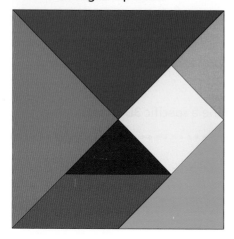

Cut a copy of the tangram into the seven separate pieces.

Task 1: Arrange the seven pieces to make two squares that are exactly the same size.

Task 2: Use the square piece and four of the triangles to make a parallelogram.

Task 3: Use some or all of the pieces to make other types of quadrilateral.

Key terms

Two shapes are **congruent** if they have exactly the same shape and size. In congruent shapes, **corresponding** sides and angles are equal.

2.1 Congruency

Geometric conventions

Capital italic letters are used to label a point or a vertex.

The line segment AB is the line joining point A to point B.

The triangle ABC is the triangle with sides AB, BC and AC.

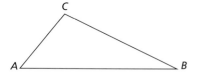

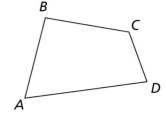

The quadrilateral $ABCD$ is the quadrilateral with sides AB, BC, CD and AD.

Tip

You should always label 2D shapes with the letters going in the same direction around the shape – usually clockwise but anticlockwise is also correct.

Note: you would always keep the letters in alphabetical order.

For example, you would not label a quadrilateral $ABDC$.

 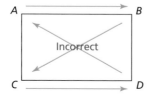

Worked example 1

Here are two congruent triangles.

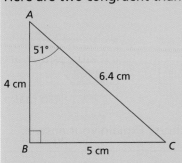 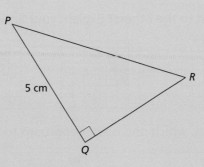

Not to scale

a) Which side in triangle *ABC* measures 6.4 cm?

b) Write down the length of side *PR.*

c) Find the size of angle *RPQ.*

a) *AC* is 6.4 cm	Identify the side by the vertices at either end.	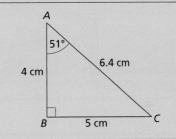
b) *PR* = 6.4 cm	*PR* is opposite the right angle. The triangles are congruent so *PR* will be the same size as the side opposite the right angle in triangle *ABC*. This is side *AC*, which is 6.4 cm.	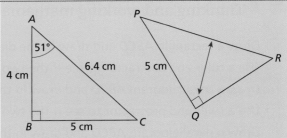
c) Angle *RPQ* = 39°	Corresponding angles in congruent triangles are equal. Angle *RPQ* is the angle between *PR*, known to be 6.4 cm and *PQ*, known to be 5 cm. This corresponds to angle *ACB*. The angles in a triangle add up to 180°, so angle *ACB* is 180° − (90° + 51°) = 39°.	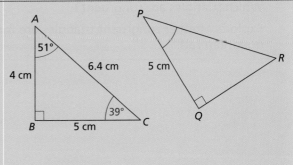

Think about

If two squares are congruent, what can you say about the sides of each square?

Discuss

How can two equilateral triangles be congruent?

1 Which shape is not congruent to the others? Explain your answer.

2 In each of these sets of shapes, which one is not congruent to the others? Explain your answer.

a)

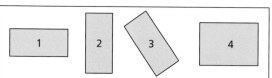

b)

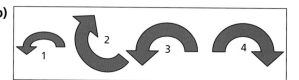

c)

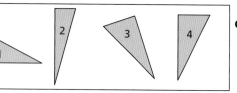

d)

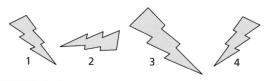

Thinking and working mathematically activity

- Draw a rectangle *ABCD* and draw in the diagonal *AC*.

  Use a ruler and protractor to see if the two triangles *ABC* and *CDA* are congruent.

- Draw a parallelogram *ABCD* and draw in the diagonal *BD*.

  Use a ruler and protractor to see if the two triangles *BCD* and *ABD* are congruent.

- Draw a parallelogram *ABCD* and draw in the diagonals *AC* and *BD*.

  Which triangles are congruent?

  Explore whether congruent triangles are formed when a diagonal is drawn on other quadrilaterals.

3 The diagram shows some triangles drawn on a dotted grid.

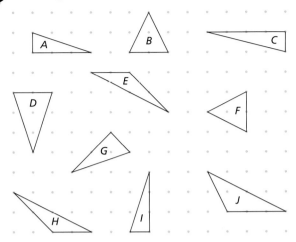

Find three pairs of congruent triangles.

4 The diagram shows a hexagon divided into 17 triangles.

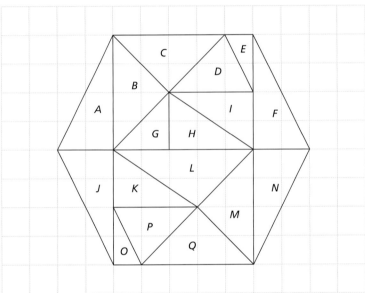

a) Write down the letters of all the triangles that are congruent to triangle C.

b) Which triangle is congruent to triangle *D*?

c) Which two triangles are congruent to triangle *H*?

5 These two triangles are congruent.

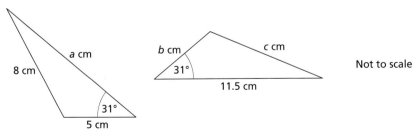

Write down the values of *a*, *b* and *c*.

6 Triangle *ABC* is congruent to triangle *LMN*.

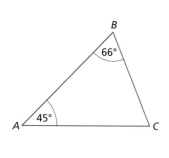 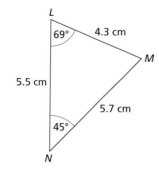

Not to scale

Find:

a) angle *LMN*

b) the length of *BC*

c) the length of *AB.*

7 **a)** Which of the following triangles is congruent to triangle *T*?

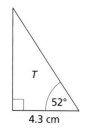 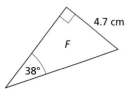 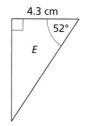

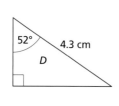 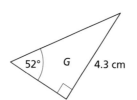

Not to scale

b) Explain why the other triangles are not congruent to triangle *T*.

8 These two quadrilaterals are congruent.

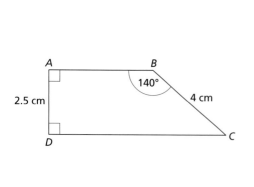 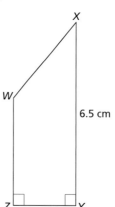

Not to scale

> **Tip**
>
> Use the fact that the angles in any quadrilateral add up to 360°.

Write down:

a) the length of *CD*

b) the length of *WX*

c) the length of *YZ*

d) the size of angle *XWZ*

e) the size of angle *BCD.*

9 Here are two congruent isosceles triangles.

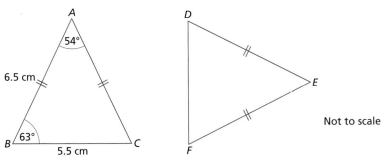

Not to scale

Write down:

a) the length of *DF*

b) the length of *DE*

c) the size of angle *FDE*

d) the size of angle *DEF*.

10 These two pentagons are congruent.

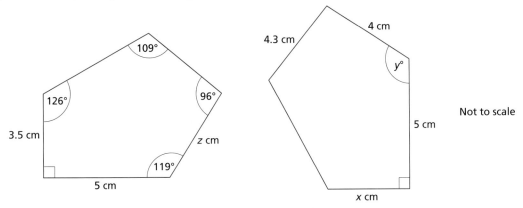

Not to scale

Find the values of *x*, *y* and *z*.

11 Channa says these two triangles are congruent.

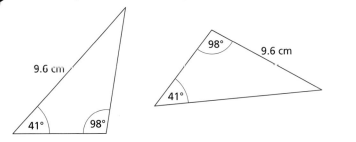

Not to scale

Is Channa correct? Explain your answer.

12 Here is a description of a quadrilateral:

Two sides measure 3 cm and two sides measure 7 cm.

Noor says that all quadrilaterals matching this description must be congruent.
Draw diagrams to show that Noor is incorrect.

2.2 Circles

Key terms

A **circle** is a 2-dimensional shape made by drawing a curve where all the points are the same distance from the centre.

The distance around the edge of a circle is called the **circumference.**

A **radius** of a circle is a straight line from its centre to any point on its circumference.

A line that connects two points on the circumference of a circle is called a **chord**.

A **diameter** is a chord that passes through the centre of the circle.

A **tangent** to a circle is a straight line that just touches the circumference at one point and does not cut through it.

Lines that meet at 90° are called **perpendicular**.

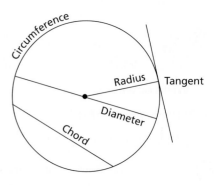

Thinking and working mathematically activity

- Draw a circle with radius 4 cm and label the centre O.
- Draw a radius and label the end at the circumference T.
- Draw a tangent to the circle at the point T. Label the tangent AB. Use a protractor to find the size of angle OTB.
- Repeat the above for a different radius. Write down what you notice.
- Do you think this will always happen? Investigate to find out.
- Do you think this will happen on any sized circle? Investigate to find out.

Think about

Can a chord ever be longer than a diameter?

Worked example 2

Give a description of each diagram below.

a)   b)   c)

a) Two circles touching each other. The small circle has a diameter the same size as the radius of the large circle.	See the bottom of the small circle just touches the circumference of the large circle. The top of the small circle just reaches up to the centre of the large circle.
b) Two tangents drawn on a circle that are perpendicular to each other.	See two tangents and notice they are at right angles to each other.
c) Three chords in a circle meeting to form a triangle.	Notice the triangle.

1 **a)** Draw a circle with radius 2.5 cm.

   b) Draw a horizontal diameter.

   c) Draw a radius that is perpendicular to the diameter.

2 **a)** Draw a circle with radius 3.2 cm.

   b) Mark four points equally spaced on the circumference.

   c) Draw chords to connect each point to each other point.

3 **a)** Draw a circle with radius 2.9 cm.

   b) Draw on a vertical chord of length 4 cm.

   c) Draw on a horizontal chord of length 4.2 cm.

Thinking and working mathematically activity

- Draw a circle with diameter 6 cm.

   Mark the ends of the diameter A and B.

   Mark a point C on the circumference.

   Measure the size of the angle ACB.

- Mark another point D on the circumference.

   Measure the size of the angle ADB.

- Repeat with different circles. Write down what you notice.

4 Give a description of each diagram below.

   The first one has been done for you.

a)

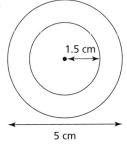

The diagram shows two circles with the same centre.

The small circle has radius 1.5 cm.

The larger circle has diameter 5 cm.

b)

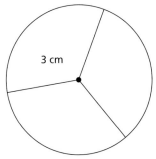

c)

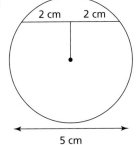

d)

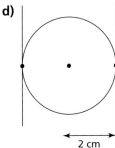

5 **Vocabulary question** Copy the sentences below and fill in the blanks using the list of words in the box. Some words are used more than once.

radius	diameter	circle	tangent
circumference	centre	chord	

a) On the edge of a _____ all the points are the same distance from the _____. The edge is called a _____.

b) A _____ of a circle is a line from the centre to the _____.

c) A straight line that connects any two points on the edge of a circle is called a _____. If this line cuts through the _____ it is called a _____.

d) A _____ is a straight line just touching a _____ at one point on its _____.

6 Write down if each of the following are always true, sometimes true or never true.

Give a reason for each answer.

a) The diameter of a circle is twice as long as the radius.

b) A chord in a circle is larger than the radius.

c) The tangent to a circle is perpendicular to a radius.

d) Two diameters of a circle will be perpendicular to each other.

2.3 3D shapes

Key terms

3D shapes have **faces**, **edges** and **vertices**.

A surface can be flat or curved but a face is always flat with straight lines as sides and is usually part of a polygon. A sphere has a curved surface. A cube has 6 square faces.

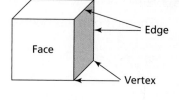

A **prism** is a 3D shape with the same **cross-section** all the way along its length.

We name a prism using the name of the shape on its cross-section. For example, triangular prism or octagonal prism.

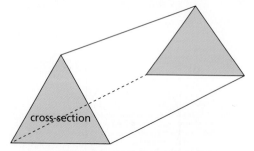

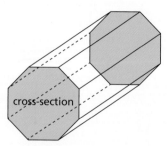

A **pyramid** is another special type of 3D shape.

Pyramids can have different shaped bases. The other faces of a pyramid are all triangles which meet at a single vertex.

For example, here is a pyramid with a base that is a regular hexagon.

We name a pyramid using the name of the shape on its base.
For example, this is a hexagonal pyramid.

A pyramid with a triangle for a base is called a **tetrahedron**, which means four-faced shape, but you could call it a triangular pyramid.

Here are some 3D shapes that have curved surfaces.

sphere　　　　**cone**　　　　**cylinder**　　　**hemisphere**

Thinking and working mathematically activity

Investigate whether the following statements are always true, sometimes true or never true.

Give reasons for your answers.

- A pyramid has an odd number of edges.
- The number of edges of a prism is a multiple of 3.
- The number of faces of a pyramid is an even number.

Worked example 3

a) Name these solids.

i) 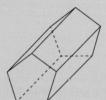　　　ii)

b) Describe the faces of a triangular prism.

c) State the number of vertices, edges and faces on a triangular prism.

a) i) This shape has an octagonal base. All of the other faces are triangles which meet at a single vertex. This is an octagonal pyramid.	✓ all faces are triangles except the base and these meet at a vertex → pyramid ✓ base is an octagon → octagonal pyramid
ii) This shape has a constant cross-section of a pentagon. So, this is a pentagonal prism.	✓ constant cross-section connected by rectangular faces → prism ✓ cross-section is a pentagon → pentagonal prism
b) Start by sketching a triangular prism. Now you can identify the different faces.	A triangular prism has five faces: two triangular faces and three rectangular faces.

c) Look at your triangular prism.

Vertices:

You can count the vertices shown by the red dots. There are 6 vertices.

Edges:

You can count the edges on the sketch of the triangular prism. There are 9 edges.

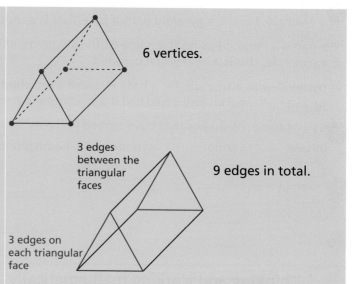

6 vertices.

3 edges between the triangular faces

9 edges in total.

3 edges on each triangular face

Exercise 3

1 Name each of these 3D shapes.

a)

b)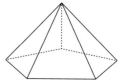

c)

d)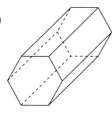

e)

f)

2 State the number of vertices, edges and faces of each of these shapes.

a) cuboid   b) triangular prism   c) tetrahedron

d) hexagonal pyramid   e) pentagonal prism

3 Describe the number and shapes of any faces of the following.

a) a cube   b) a hexagonal prism   c) a square-based pyramid

d) a cuboid   e) an octagonal pyramid

4 A shape has 5 faces.

Write down the letters of the shapes it could be.

A:	B:	C:	D:	E:
square-based pyramid	pentagonal pyramid	pentagonal prism	tetrahedron	triangular prism

5 Which of these shapes does not have a curved surface?

A: cone   **B:** hemisphere   **C:** pyramid   **D:** cylinder   **E:** sphere

6 Explain the difference between:

a) a cube and a cuboid   **b)** a pyramid and a prism   **c)** a cone and a pyramid

7 Paloma says, 'A cube is a prism'.

Do you agree with Paloma? Explain your answer.

8 Dylan has a shape with 6 faces.

Bob says it **must** be a cuboid.

Explain why Bob could be incorrect.

9 Look at these three pyramids.

 P   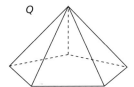 Q   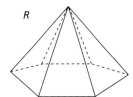 R

a) Copy and complete the table below.

	Shape of base	Number of vertices	Number of faces
P			
Q			
R			

b) Write down what you notice about each pyramid.

c) Do you think this will happen in any shaped pyramid? Give a reason for your answer.

Consolidation exercise

1 State whether or not the following pairs of triangles are congruent.
Give reasons for your answers.

a)

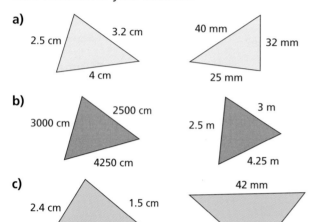

b)

c)

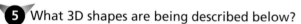

2 A star is drawn inside a regular pentagon.

a) How many triangles are formed?

b) Name the triangles which are congruent to each other.

3 **a)** Draw a circle with radius 3.3 cm.

b) Draw a diameter AB.

c) Draw tangents to the circle at points A and B.

d) What can you say about both tangents?

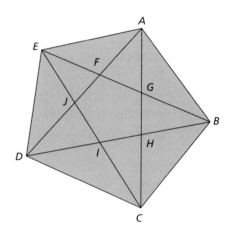

4 A pyramid has seven faces. How many vertices does it have?

5 What 3D shapes are being described below?

a) This shape has four faces and four vertices.

b) This shape has one face and no vertices.

c) This shape has seven faces, 15 edges and 10 vertices.

d) This shape has two faces that are octagons and eight faces that are rectangles.

6 The edges of a tetrahedron $ABCD$ are all the same length.

a) Draw a sketch of the tetrahedron.

b) Name the congruent triangles found on the tetrahedron.

7 Six of the faces of a 3D shape are congruent to each other.

The other two faces of this shape are also congruent to each other.

Write down the name of this shape.

End of chapter reflection

You should know that...	You should be able to...	Such as...
• Corresponding sides and angles are equal in congruent shapes.	Find side lengths and angles in congruent shapes by identifying corresponding sides.	The two quadrilaterals shown are congruent. What is the length of AB and what is the size of angle ADC?
• Lines that are at right angles to another are **perpendicular**.	Recognise perpendicular lines.	These two lines are perpendicular.
• Parts of a circle can be labelled as a radius, a diameter, the circumference and a chord. • A tangent is a line just touching the circumference of a circle once only.	Label the different parts of a circle. Draw a tangent to a circle.	Name each of these parts of a circle.
• 3D shapes have faces, edges and vertices. • A prism is a 3D shape with a constant cross-section. • Pyramids can have different shaped bases. The other faces are all triangles meeting at a single vertex.	Name common 3D shapes. State the number of faces and edges of a 3D shape. Describe the faces of a 3D shape.	a) Name these 3D shapes. b) State the number of faces, vertices and edges of a pentagonal prism. c) Describe the faces of a square based pyramid.

3 Collecting data

You will learn how to:

- Select and trial data collection and sampling methods to investigate predictions for a set of related statistical questions, considering what data to collect (categorical, discrete and continuous data).
- Understand the effect of sample size on data collection and analysis.

Starting point

Do you remember...

- planning a statistical investigation?

  For example, planning an investigation to see if there is a connection between hand span and height.

- making predictions about what you think your data will show?

  For example, predicting that taller people will have a larger hand span than shorter people.

- collecting data to see if your prediction is correct?

  For example, collecting data from 15 students in your class.

This will also be helpful when...

- you learn more about the problems linked with some methods for collecting data
- you learn more about presenting your data in different ways
- you learn more about calculating averages from data to help analyse it.

3.0 Getting started

A person's pulse rate is the number of heartbeats per minute.

Pulse rates are affected by a person's age and weight, health, medication and even air temperature and body position. A normal pulse rate for a person older than 10 years old is between 60 and 100 beats per minute when they are at rest.

Predict what you think might happen to your pulse rate when you exercise.

Let's find out!

Measure your pulse rate when you are sitting down.

Jog on the spot for one minute, then measure your pulse rate again. Record the number of beats per minute immediately after you stop jogging.

- What happened to your pulse rate?
- Compare your results to others in your class – what do you notice?
- To investigate further, what other information could you collect?

> **Tip**
>
> To measure your pulse rate, put your fingers on your wrist and count the beats. Record the number of beats per minute.

> **Did you know?**
>
> You can also find your pulse in your neck. Place your index and middle fingers on your neck to the side of your windpipe.

3.1 Types of data

Worked example 1

Maxine is going to compare some flowers in her local park.
She is going to be comparing:

- the number of petals
- the height of the flower
- the colour of the petals
- the name of the flower
- the diameter of the stem.

Categorise these types of data under the headings: categorical data, discrete data and continuous data.

Categorical data the name of the flower the colour of the petals	The name and the colour cannot be counted or measured.
Discrete data the number of petals	The number of petals can be counted and the results will be given as whole numbers.
Continuous data the height of the flower the diameter of the stem	The height and diameter are measurements and will be rounded to perhaps 1 or 2 decimal places.

Key terms

Data types:

Categorical data are data that can be put into categories. Examples of categorical data are country of birth, favourite subject, name and eye colour.

Discrete data are numerical values which can be recorded exactly. Discrete data usually is counted. Examples of discrete data are number of brothers and sisters, number of books and number of employees.

Continuous data needs to be measured and then rounded to a suitable degree of accuracy. Examples of continuous data are height, mass and speed.

Exercise 1

1 Tim is collecting information from people about their use of mobile phones.

Decide if each of the following variables is categorical, discrete or continuous data.

a) the time spent using their mobile phone yesterday

b) the number of text messages they sent yesterday

c) the make of their mobile phone.

2 Sofia is measuring the height of sunflowers in her garden. Is the height of the sunflowers discrete or continuous data?

3 Humna is collecting data about the voluntary work that some of her friends and family do. Decide if each of the following variables is categorical, discrete or continuous data.

- time spent per week doing voluntary work
- age (in years)
- number of volunteers in their organisation
- gender (male or female)
- type of voluntary work undertaken.

4 A football coach wants to find out about the fitness levels and performance of his players.

He collects the following data from each player:

- heart rate (beats per minute)
- height
- mass
- age (in years)
- time taken to run a 100 metre sprint
- number of goals scored in training.

Write down whether each variable is continuous data or discrete data.

5 Robin wants to investigate the shopping habits of his classmates. List two examples of discrete data he could collect and two examples of continuous data he could collect.

6 Brigita says that age is discrete data. Jordan says that age is continuous data. Explain how Brigita and Jordan can both be correct.

Thinking and working mathematically activity

Copy this table.

Categorical	Discrete	Continuous

Imagine that you want to collect data from people about music.

Think of examples of variables that you could collect data about. Write your examples in the correct column in the table. Try to find at least two variables for each column.

3.2 Data collection methods

Key terms

When you investigate a statistical question, you may need to collect your own data.

There are several common methods for collecting data:

1 **Observation** – This is when you collect data by observing or counting. For example, you could count the number of cars that pass along a road.

The collected data can be recorded in a **data collection sheet** which is a type of table.

2 **Interviews** – This is when you collect data by asking people questions. For example, you could ask people about their views on how to protect the environment.

A **focus group** is a group of people who are brought together to discuss their views on a particular subject.

3 Questionnaires – A questionnaire is a set of questions that you send out for people to complete. You can design a questionnaire for people to complete online or on paper.
For example, you could prepare a questionnaire to ask people about their holidays.

Questions in a questionnaire should be clear and easy to answer. A questionnaire can include yes/no answers, tick boxes, numbered responses, word responses and questions requiring a longer written answer.

A question is said to be **biased** if it is worded such that a particular answer is favoured over other answers. To get reliable data from a questionnaire, the questions should be **unbiased**.

Worked example 2

A website manager wants to find out about the age and gender of people who use the site and how often they use it. Design a short questionnaire for users to complete to collect this data.

The manager wants to find out about the age and gender of people who use the site and how often they use it.	The manager wants to know three things: **1)** age **2)** gender **3)** how often people use the site.
1) What is your age? • Under 18 ☐ • 18–30 ☐ • 31–60 ☐ • 61 or over ☐	The first question is about age so you can either ask for an exact age or provide categories. Categories are easier to analyse. Make sure that the options do not overlap and that every possible age is included.
2) What is your gender? • Female ☐ • Male ☐	The second question is about gender, so you can ask a simple male or female question here.
3) How often do you use this site **per week**? • 0 or 1 times ☐ • 2 or 3 times ☐ • 4–10 times ☐ • more than 10 times ☐	For the third question you can use option boxes to make the data easier to analyse. Make sure the options do not overlap and that every possible answer is included. Make sure you give a time period for people to base their answer on, for example, per month or per week.

Exercise 2

1 Mo has been asked to explore the question:

Do adults sleep more than children?

He has decided to collect data from people in a local town centre.

Which two of these things will it be most useful for him to collect data about?

a) gender of person

b) age of person

c) number of hours slept last night

d) time they woke up this morning.

2 A café owner is trying to find out which drinks are its bestsellers.

She will record the drinks that customers buy one morning. Copy and complete the data collection sheet. Fill in the first column with names of possible drinks.

Drink	Tally	Frequency
....		
....		
....		
....		
....		
....		

3 Valeria is researching students' favourite subjects.

She completes a survey of students and asks them, 'What is your favourite subject?'

Design a table for Valeria to use to record her results.

4 Write a questionnaire question that could be used to collect data in each of these situations. Include some response boxes for each of your questions.

a) Moira wants to find out how many hot drinks people have in a day.

b) Amol wants to find out how far (in kilometres) students in his school travel to school.

5 A mobile phone company believes that young people use their phones more than older people. Write a questionnaire of at least three questions to help the company collect data to test their theory. Try to include response boxes for some of the questions.

6 Jaspreet wants to collect information about how many times people play sport.

Here is the question she has written:

How many times do you play sport?

0 to 2 ☐   2 to 4 ☐

4 to 6 ☐   more than 6 ☐

a) Identify two problems with this question.

b) Write a suitable question for Jaspreet to use.

7 Rewrite the following questions so they are not biased.

a) Do you agree that the environment is important?

b) The business has made a profit for the last five years. Do you think the business is successful?

c) Are you in favour of opening a new cinema to give young people more to do?

Thinking and working mathematically activity

Design a questionnaire for collecting data about music. Write questions that will help you to collect information about things like:

- age and gender
- the type of music people like
- how people listen to music (radio, streaming, …)
- the amount of time people listened to music yesterday.

Include some tick boxes for each of your questions.

Give your questionnaire to a few friends to complete. Ask them to tell you how easy it was to complete. Use their comments to make some improvements to your questionnaire.

8 Jana wants to know how many passengers there are in cars using a specific road. Suggest a way that Jana can collect suitable data.

9 Lucasz is opening a book shop in a town. He wants to find out whether people in the town would use his shop. Suggest a way that Lucasz can collect suitable data.

10 A chocolate company has developed a new type of chocolate bar. The company wants to find out the opinions of people about this chocolate bar before it is sold in the shops. Suggest a way that the company can get this information.

11 Geraldine is investigating the average height of people in different countries around the world. Which would be the most sensible way for her to collect useful data? Explain your answer.

  a) measure her own height

  b) measure the height of people in her class

  c) look on the internet

  d) phone up people in different cities and ask them for their height.

12 Elliot is researching what people think about libraries. He designs a survey to complete. Which would be the best group of people for him to survey? Explain your answer.

  a) people in his local library

  b) children at a school

  c) people in a town centre

  d) people in a local bookshop.

3.3 Choosing a sample

Key terms

The set of all people or things you want to find out about is called the **population**.

You may not want (or be able) to collect information from everybody in the population. Instead, you can collect information from a **sample**. A sample is a selection of people from the population from which data is obtained. The sample should be **representative** of the population – this means that the sample should be as similar as possible to the whole population.

A sample is called a **random sample** if every member of the population has an equal chance of being picked. An easy way to get a random sample is to pick names from a hat. Another way is to number everyone in the population and then pick the sample using a random number generator.

The **sample size** is the number of people you decide to survey from the total population.

Worked example 3

A school has 1000 students. The head teacher wants to investigate the average time that students at her school spend travelling to school.

a) Write down the population for the head teacher's investigation.

b) She decides to ask a sample of ten students in Year 7 for their journey time to school.

Write down two comments about her sample.

c) Explain how the head teacher could choose the students who will be in her sample.

a) The population is all students in the school.	The population is all of the people that she wants to find out information about.
b) Selecting ten students is too small a sample size to gain a variety of results.	The more students you survey, the more reliable your results will be. In this instance it may be more appropriate to ask about 100 students.
The head teacher's sample will not be representative of all students in the school as she is only asking students from Year 7.	A selection of students from all year groups should be used, not just students from one year group.
c) She could put the names of all the students into a hat and pick out names without looking.	She should choose her sample randomly. This will help to make sure the students in her sample are representative of those in the whole school.

Exercise 3

1 A gym has 500 members. The manager wants to select a sample of gym members to ask them what they think about the gym.

a) Explain why asking a sample of five gym members may not give reliable results.

b) Suggest a more suitable sample size.

2 Pippa wants to select a sample of people who work at her office. She has a list of all 100 people who work at the office. She selects the first person on the list and then every fifth person.

How many people does Pippa select?

3 Decide whether each statement below is true or false.

a) A very small sample is likely to be representative of the whole population.

b) Data from a large sample takes longer to analyse than data from a small sample.

c) Results from a large sample are likely to be more reliable than results from a small sample.

> **Discuss**
>
> Why in Worked example 3 why might the journey times to school for Year 7 students be different to the journey times for older students?

> **Did you know?**
>
> A sample that is chosen by selecting people at regular intervals is called a **systematic sample**.

4 Mathias wants to find out how often people from his town go to the cinema. He decides to interview people for his survey.

    a) What is the population for his survey?

    b) Give a reason why Mathias is unlikely to want to collect information from everyone in the population.

    c) Mathias decides to interview 50 people from the town. He plans to conduct interviews by asking people in the street. Suggest a suitable time and place where Mathias could conduct the survey.

5 Amy wants to investigate how happy residents living in an apartment block are with their apartments. She decides to select a sample of residents.

    a) Write down the population for her investigation.

    b) Amy decides to ask all the residents living on the first floor. Explain why her sample is not a random sample of residents living in the block.

    c) Explain how Amy could obtain a random sample of residents.

6 Andrei wants to know how many hours per week people spend exercising. He asks 30 people at his local running club.

    Give one disadvantage of this approach.

7 Jeremy has been asked to investigate the ages of people who use the library. He stands in the library between 10 a.m. and 11 a.m. on a Wednesday and he tallies the number of people who come in during that period.

    Give two disadvantages of this method.

8 A college has 2000 students and 100 staff. The college wants to find out what students and staff think about a new timetable. Write down a plan for data collection. Ensure that you include all of the details.

> **Tip**
>
> In question 8, include information like:
> - how many students and how many staff the college should question
> - how these students and staff could be selected
> - how the college could collect information from these students.

Thinking and working mathematically activity

Use the questionnaire that you designed in section 3.2 to collect some data about music from students in your school.

Think about:
- What do you expect to find out? (These are your conjectures.)
- Who do you want to collect data from? Who is your target population?
- How many people will you give your questionnaire to?
- How will you select the people you will ask to complete your questionnaire?

Explain why your choices will lead to reliable data being collected.

After you have collected your data, make some conclusions. Were your conjectures shown to be true?

Consolidation exercise

1 Copy this table onto a piece of paper.

True statements	False statements

Write these statements in the correct place in the table:

- number of pets is continuous data
- area is continuous data
- favourite TV program is categorical data
- speed is discrete data.

2 There are 500 nurses who work at a hospital. Jill wants to find out what they think about a new work rota. She decides to ask five nurses for their opinion. Comment on Jill's sample size.

3 George owns a fruit shop. He wants to find out what type of fruit his customers like best.

a) What is the population for his investigation?

b) George decides to interview 30 of his customers. Suggest one way he can select these customers.

4 Emily designs a questionnaire to find out about Mathematics homework. Here is her questionnaire:

1 Do you agree that Mathematics homework is very useful?

- Yes
- No

2 How long do you spend doing your Mathematics homework per week?

- 0–30 minutes
- 30–60 minutes
- 60–90 minutes

3 Where do you usually complete your homework?

- In the library
- At home

4 What do you think of your homework?

- Always interesting
- Sometimes interesting
- Always boring

> **Tip**
>
> Try answering the questionnaire yourself to see what the issues are.

Give feedback on Emily's questionnaire and make suggestions about how she could improve it.

5 Sophie wants to find out what people think about government spending on roads.
Which of these questions is the best? Explain your answer.

A: Do you think you pay too much tax?

B: Do you think our roads are good?

C: Would you be prepared to pay more tax if the money raised was spent on improving roads?

D: Do you think the government spends too much money on themselves and not enough on roads?

6 **Vocabulary question** Copy and compete the sentences below and fill in the blanks using the list of words in the box.

categorical	exact	sample size	discrete
measurements	survey	continuous	results

When conducting a _____ it is important that your _____ is large enough to give reliable _____.

Sometimes the data you collect will be _____, such as eye colour or favourite sport.

Other data is numerical. _____ data is counted data and is _____ values, such as the number of students in a class. _____ data (for example height or mass) are _____ and cannot be written down exactly.

End of chapter reflection

You should know that...	You should be able to...	Such as...
There are different types of data.	Identify data that is categorical, continuous and discrete.	Which of the following variables are continuous? height, number of siblings, mass, number of press ups
There are different ways of collecting data such as questionnaires, face-to-face interviews or by observation.	Identify which information needs to be collected. Decide which method of data collection is appropriate.	James is going to investigate whether adults eat more or less chocolate than teenagers. a) What information should he collect? b) How could James collect the information he needs?
Questions in a questionnaire should be clear and simple to answer.	Design a questionnaire to find out relevant data.	Design a questionnaire to find out how much money people spend on chocolate.
It is usually simpler to collect data from a sample rather than from everyone in the population.	Decide on a suitable way to choose a sample.	Serena wants to choose a sample of 20 houses from the 200 houses on a road. How could Serena choose her sample?
A larger sample size gives more reliable results.	Choose an appropriate sample size for the population you are investigating.	Explain why a sample of 10 from a population of 1000 may be an inappropriate size.

4 Negative numbers and indices

You will learn how to:
- Estimate, add and subtract integers, recognising generalisations.
- Estimate, multiply and divide integers including where one integer is negative.
- Understand the relationship between squares and corresponding square roots, and cubes and corresponding cube roots.

Starting point

Do you remember...

- how to add and subtract integers where one integer is negative?

  For example, what is $-3 + 8$?
- how to estimate the answer to an addition or subtraction of integers?

  For example, estimate the answer to $81 - 32$.
- the square numbers between 1 and 100?

  For example, which of these are square numbers: 1, 2, 10, 36, 50, 64, 84?
- the cube numbers between 1 and 125?

  For example, which of these are cube numbers: 1, 3, 8, 27, 32, 64, 75?

This will also be helpful when...

- you learn to manipulate algebraic expressions involving negative terms
- you learn to recognise squares of negative numbers
- you learn to use higher indices than squares and cubes.

4.0 Getting started

Use a calculator to explore the four operations with positive and negative numbers.

Work systematically and write down your findings. The table shows some ideas you could use.

Investigate	Examples	Questions to think about
Adding a negative number to another number	$6 + (-3)$	Is $6 + (-3)$ the same as $6 - 3$, $6 + 3$, $-3 + 6$ or $-3 - 6$?
	$-4 + 5$	Can you write $-4 + 5$ in another way?
	$-4 + (-5)$	Is $-4 + (-5)$ the same as $-4 - 5$, $-4 + 5$, $-5 + 4$ or $-5 - 4$?
Subtracting a negative number from another number	$7 - (-3)$	Is $7 - (-3)$ the same as $7 + 3$, $-3 + 7$ or $-3 - 7$?
	$-7 - (-3)$	Can you write $-7 - (-3)$ in another way?
Multiplying a positive number by a negative number	$3 \times (-5)$	Does -5×3 give the same answer as $3 \times (-5)$?
Dividing a positive number by a negative number	$8 \div (-2)$	How is the answer different from the answer to $8 \div 2$?
Dividing a negative number by a positive number	$-8 \div 2$	Can you find other integer divisions that give the same answer?

4.1 Adding and subtracting integers

Key terms

Integers are whole numbers, including zero (for example, −340, −5, 0, 11, 2019).

Positive numbers are numbers greater than zero (for example, 1000, 8, 0.1, $\frac{2}{3}$).

Negative numbers are numbers less than zero (for example, −1000, −8, −0.1, −$\frac{2}{3}$).

The four operations are multiplication, division, addition and subtraction.

The **sum** of two or more numbers is the answer when you add them.
For example, the sum of 4 and 5 is 9.

The **difference** between two numbers is the answer when you subtract one number from the other.
For example, the difference between 10 and 3 is 7.

Did you know?

Over two thousand years ago, Chinese mathematicians used rods to represent numbers. They used

red rods for positive numbers and black rods for negative numbers. For example, ▐▐▐▐ represented 4

if the rods were red, and −4 if the rods were black. ⊤ represented 6 if the rods were red and −6 if the rods were black.

In many countries today, however, the opposite colours are used when writing amounts of money. Red represents a negative amount (a debt) and black represents a positive amount (a credit). In English, the phrase 'in the red' means 'in debt'.

Worked example 1

Find:

a) 5 + (−3)   **b)** −2 − (+3)   **c)** 3 − (−2)

a) 5 + (−3) = 5 − 3 = 2	5 + (−3) means 'positive 5 plus negative 3'. Adding a negative number is the same as subtracting a positive number.	Three of the +1s and the three −1s make 0, leaving +2. 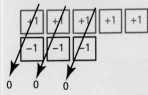

b) $-2 - (+3) = -2 - 3$ $\qquad = -5$	$-2 - (+3)$ means 'negative 2 minus positive 3'. $-2 - (+3)$ is the same as $-2 - 3$	There are no +1s in -2. But it is always possible to add $(+1\ -1)$ pairs, since these equal 0. Add three $(+1\ -1)$ pairs to -2. 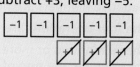 Now you can subtract +3, leaving -5.
c) $3 - (-2) = 3 + 2$ $\qquad = 5$	$3 - (-2)$ means 'positive 3 minus negative 2'. Subtracting a negative number is the same as adding a positive number.	There are no -1s in 3. Add two $(+1\ -1)$ pairs to 3. 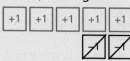 Now you can subtract -2, leaving +5. 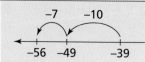

Worked example 2

Find:

a) $-18 + 23$ **b)** $-39 - 17$

a) $-18 + 23 = 5$	Reversing the order of the numbers makes this calculation easier: $23 - 18 = 5$. Alternatively, sketch a number line and do the calculation in shorter 'jumps'.	
b) $-39 - 17 = -56$	Start at -39. Move 17 along the number line in the negative direction.	

Exercise 1   1–8

1 Copy and fill in the missing numbers.

a) 8 °C warmer than −1 °C is _____ °C.   b) 5 °C warmer than −12 °C is _____ °C.

c) 15 °C colder than 3 °C is _____ °C.   d) 7 °C colder than −6 °C is _____ °C.

e) _____ °C warmer than −10 °C is −4 °C.   f) 12 °C colder than _____ °C is −2 °C.

g) _____ °C colder than −3 °C is −14 °C.   h) 30 °C warmer than _____ °C is 21 °C.

2 Copy and write the missing numbers.

$$7 - 2 = 5$$
$$7 - 1 = \underline{\hspace{1.5cm}}$$
$$7 - 0 = \underline{\hspace{1.5cm}}$$
$$7 - (-1) = \underline{\hspace{1.5cm}}$$
$$7 - (-2) = \underline{\hspace{1.5cm}}$$
$$7 - (-3) = \underline{\hspace{1.5cm}}$$

Describe the pattern.

Discuss

These calculations are incorrect. What mistakes have been made?

$5 - 7 = 2$

$-5 - 4 = 9$

$5 - (-2) = -7$

3 Copy and find the missing numbers.

a) $-12 + 4 = \underline{\hspace{1.5cm}}$

b) $-12 - 4 = \underline{\hspace{1.5cm}}$

c) $\underline{\hspace{1.5cm}} + 22 = 3$

d) $\underline{\hspace{1.5cm}} + 22 = -3$

e) $-5 + \underline{\hspace{1.5cm}} = 17$

f) $-5 - \underline{\hspace{1.5cm}} = -17$

g) $11 - 11 = \underline{\hspace{1.5cm}}$

h) $42 - 60 = \underline{\hspace{1.5cm}}$

i) $-25 + 52 = \underline{\hspace{1.5cm}}$

j) $\underline{\hspace{1.5cm}} + 8 = 0$

4 How many pairs of integers between -20 and 20 can you find with a sum of 7? How do you know you have found them all?

5 Find:

a) $5 - (+3)$

b) $7 + (-4)$

c) $-3 + (-2)$

d) $4 - (-8)$

e) $-6 + (-7)$

f) $5 - (-8)$

g) $-2 - (-6)$

h) $7 - (+9)$

i) $9 - (-9)$

j) $-5 + (-9)$

k) $-8 - (-9)$

l) $-7 + (-2)$

Think about

Where do you see negative numbers in real-life? When do we add or subtract positive and negative numbers in real-life?

6 Write your own addition and subtraction questions that have the answers below.

Only use the numbers 8, -10 and -5.

a) 3

b) -2

c) -18

d) 5

7 In a magic square, all the rows, columns and diagonals add to the same total.

Copy and complete these magic squares.

Tip

First, find the sum of a complete row or diagonal.

	-5	
-4	7	0

2		
	-5	6
		-12

8 Explain, using examples, whether each statement is always true, sometimes true, or never true.

a) When you **add** a **positive** integer to a **positive** integer, the answer is **positive**.

b) When you **add** a **negative** integer to a **negative** integer, the answer is **negative**.

c) When you **subtract** a **positive** integer from a **positive** integer, the answer is **positive**.

d) When you **subtract** a **negative** integer from a **positive** integer, the answer is **positive**.

e) When you **subtract** a **positive** integer from a **negative** integer, the answer is **negative**.

f) When you **subtract** a **negative** integer from a **negative** integer, the answer is **negative**.

9 Use a calculator to find:

a) 145 − (−254) **b)** −829 + (−764) **c)** 45 − (−199)

Thinking and working mathematically activity

Draw a pyramid like the one below.

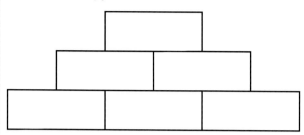

In each box on the bottom row, write a positive integer or a negative integer.

In the rows above, write in each box the sum of the two numbers below it.

Is the top number of your pyramid positive or negative?

Try to create an addition pyramid whose top number has the opposite sign from the bottom middle number.

Investigate addition pyramids further. Try to find a rule that relates the sign of the top number to the values of the bottom numbers. When is the top number zero?

4.2 Multiplying and dividing integers

Worked example 3

Find:

a) 2 × (−6) **b)** −15 ÷ 3 **c)** 12 ÷ (−4)

a) 2 × (−6) = −12 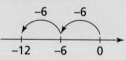	Work out 2 lots of −6. (−2 × 6 also equals −12. Can you see why?)	+ × − = − − × + = −
b) −15 ÷ 3 = −5	How many threes make −15? 3 × (−5) = −15 so − 15 ÷ 3 = −5 A negative number divided by a positive number equals a negative number.	− ÷ + = −
c) 12 ÷ (− 4) = −3	A positive number divided by a negative number equals a negative number.	+ ÷ − = −

1 The temperature changes by −2 °C every hour. What is the total change in temperature after 7 hours?

2 Find:

a) −4 × 4 b) 5 × (−7) c) −5 × 9 d) −10 × 2

e) −3 × 6 f) 7 × 5 g) 8 × (−4) h) 5 × (−11)

3 Find:

a) −4 ÷ 4 b) 20 ÷ (−2) c) −25 ÷ 5 d) −42 ÷ 7

e) 32 ÷ 8 f) 45 ÷ (−9) g) 8 ÷ (−4) h) −28 ÷ 4

4 Write true or false for each calculation. If it is false, correct it.

a) −49 ÷ 7 = 7 b) −7 × 4 = −21 c) 80 ÷ (−10) = 8 d) 3 × (−2) = −6

5 Write two multiplication calculations and two division calculations that give each answer.

a) −10 b) 32 c) −12

6 Find the missing numbers.

a) ____ × (−4) = −28 b) 6 × ____ = 30 c) 22 ÷ ____ = −11

d) ____ ÷ (−5) = −6 e) 9 × ____ = −63 f) ____ ÷ 4 = −6

7 You are given the result −288 ÷ 24 = −12. Without doing any working, write at least six other calculations that must be true.

8 Below are two puzzles. Copy each puzzle and write numbers in the four boxes to make all of the calculations correct.

a)
$$\Box \times \Box = -8$$
$$\times \quad \div$$
$$\Box \times \Box = 6$$
$$= 6 \quad = -2$$

b)
$$\Box \div \Box = 6$$
$$\div \quad \times$$
$$\Box \div \Box = -2$$
$$= -3 \quad = 25$$

> **Tip**
>
> In question 7, can you write another division and a multiplication? If you change the signs on numbers, can you write more multiplications and divisions?

9 Use a calculator to find:

a) 38 × (−69) b) −87 × 6 c) 4050 ÷ (−50)

d) 9126 ÷ (−3) e) −189 ÷ 3 f) −341 × 214

Thinking and working mathematically activity

x and y are integers. Look for values of x and y that give each of the types of result below.

- $x \div y$ is less than both x and y.
- $x \div y$ is greater than y.
- $x \div y$ is greater than x.
- $x \div y$ is greater than both x and y.

Write rules to summarise your findings. Try to explain why each rule works.

4.3 Estimating

Key terms

An **exact** value is the true value of a calculation. For example, the exact value of 99 × 3 is 297.

A result that is **approximate** is not exactly correct. For example, an approximate value of 99 × 3 is 300. The symbol ≈ means 'approximately equals'. For example, 99 × 3 ≈ 300.

To **round** a number is to write its approximate value. For example, 388 rounded to the nearest 10 is 390, and 388 rounded to the nearest 100 is 400.

To **estimate** is to find an approximate answer to a calculation. For example, an estimated answer to 99 × 3 is 100 × 3 = 300.

Worked example 4

Estimate:

a) 58 – 81   b) –289 – 324   c) 518 × (–42)   d) 208 ÷ (–11)

Then use a calculator to find the exact answers.

a) 58 – 81 ≈ 60 – 80 = –20 Exact answer = –23	Round each number so that you can do the calculation mentally or using a sketch. Round 2-digit numbers to the nearest 10. 58 lies between 50 and 60, and it is closer to 60. 81 lies between 80 and 90, and it is closer to 80. Find 60 – 80.	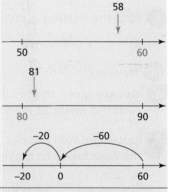
b) –289 – 324 ≈ – 300 – 300 = –600 Exact answer = –613	Round 3-digit numbers to the nearest 100. –289 lies between –300 and –200, and it is closer to –300. –324 lies between –400 and –300, and it is closer to –300. Find –300 – 300.	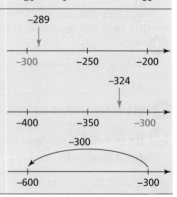

c) $518 \times (-42) \approx 500 \times (-40)$ $= -20\,000$ Exact answer $= -21\,756$	518 lies between 500 and 600, and it is closer to 500. -42 lies between -50 and -40, and it is closer to -40. If you are not sure how to find $500 \times (-40)$, rewrite it like this: $500 \times (-40) = 5 \times 100 \times (-4) \times 10$ $= 5 \times (-4) \times 100 \times 10$ $= -20 \times 1000$ $= -20\,000$	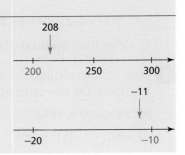
d) $208 \div (-11) \approx 200 \div (-10)$ $= -20$	208 lies between 200 and 300, and it is closer to 200. -11 lies between -20 and -10, and it is closer to -10. Find $200 \div (-10)$.	

Exercise 3

1 By rounding each number to the nearest 10, estimate:

a) $39 + 92$
b) $27 - (-51)$
c) $-94 - 22$
d) $38 - (+40)$

2 By rounding each 2-digit number to the nearest 10, estimate:

a) 8×62
b) $19 \times (-37)$
c) -93×29
d) $51 \times (-48)$

3 By rounding each 2-, 3- or 4-digit number to the nearest 10, 100 or 1000 respectively, estimate:

a) $-581 \div 10$
b) $302 \div 107$
c) $399 \div (-4)$
d) $-1072 \div 98$

4 By rounding each number to the nearest 100, estimate:

a) $198 - 307$
b) $-520 + 879$
c) $-477 - 406$
d) $621 - 888$

5 By rounding each 2- or 3-digit number to the nearest 10 or 100 respectively, estimate:

a) -22×503
b) 70×288
c) -98×909
d) $43 \times (-393)$

Think about

When is estimation useful in real-life? When is estimation inappropriate, or even dangerous?

Tip

Round the numbers so that you only need to use times table knowledge and multiplication by 10.

6 Jon wants to estimate $328 \times (-211)$. His working is below.

$328 \times (-211) \approx 330 \times (-210)$

$$
\begin{array}{r}
330 \\
\times 210 \\
\hline
000 \\
3300 \\
66000 \\
\hline
69300
\end{array}
$$

$330 \times (-210) = -69\,300$

Criticise Jon's method of estimation and write an easier method.

7 Four of the calculations below are incorrect. Use estimation to show which calculations are incorrect. Do not calculate the exact answers.

a) $38 \times (-9) = -342$ b) $78 - 83 = 161$ c) $-788 \div 81 = -97$

d) $-278 - (-411) = 689$ e) $415 \div (-39) = -106$ f) $-903 \times 28 = -25\,284$

8 a) Write three additions with estimated answer -50.

b) Write three subtractions with estimated answer -50.

c) Write three multiplications with estimated answer -200.

d) Write three divisions with estimated answer -10.

Thinking and working mathematically activity

Sometimes the estimated answer to a calculation is very close to the exact answer. Sometimes it is not so close.

- Choose one of the four operations: addition, subtraction, multiplication or division.
- Using 2-digit numbers only, find calculations where the estimated answer is close to the exact answer. Can you find calculations where the estimated answer equals the exact answer?
- Using 2-digit numbers only, find calculations where the estimated answer is not so close to the exact answer.
- Find the calculation, or calculations, where the answer is furthest from the exact answer. Try to explain why.

Choose a different operation and repeat the steps above.

4.4 Indices

Key terms

Squaring a number means multiplying the number by itself. The result is the **square** of the number. For example, the square of 5 (also called '5 squared' and written 5^2) is $5^2 = 5 \times 5 = 25$.

The **square root** of a number squares to make the number. For example, the square root of 9 (also called 'root 9') is 3 because $3 \times 3 = 9$. The symbol for square root is $\sqrt{\ }$. For example, the square root of 9 is written $\sqrt{9}$. Square roots do not have to be integers.

A **square number** (or perfect square) is the product of two copies of an integer. For example, 36 is a square number because 6 × 6 = 36.

Cubing a number means multiplying three copies of the number together. The result is the **cube** of the number. For example, the cube of 2 (also called '2 cubed' and written 2^3) is $2^3 = 2 \times 2 \times 2 = 8$.

The **cube root** of a number cubes to make the number. For example, the cube root of 27 is 3 because 3 × 3 × 3 = 27. The symbol for cube root is $\sqrt[3]{}$. For example, the cube root of 27 can be written $\sqrt[3]{27}$. Cube roots do not have to be integers.

A **cube number** (or perfect cube) is the product of three copies of an integer. For example, 125 is a cube number because 5 × 5 × 5 = 125.

An **index** or **power** shows you how many copies of a number to multiply together. It is written small, to the upper right of the number. For example, in 10^3 the index is 3, and it tells you to multiply three 10s together, or to cube 10. The plural of index is **indices**.

> Did you know?
>
> Squaring is so called because it is the same as the method for finding the area of a square: multiply a number (the side length) by itself. Why do you think cubing is so called?

Worked example 5

Write the value of:

a) 4^2 b) 5^3 c) $\sqrt{81}$ d) $\sqrt[3]{64}$

a) $4^2 = 4 \times 4$ $= 16$	4^2 means '4 squared', or 'multiply 4 by itself'.	4^2 is the area of a square of side length 4.
b) $5^3 = 5 \times 5 \times 5$ $= 125$	5^3 means '5 cubed', or 'the product of three copies of 5'.	5^3 is the volume of a cube of edge length 5.

c) $\sqrt{81} = 9$	$\sqrt{81}$ means 'the square root of 81', or 'what number multiplies by itself to make 81?' Use knowledge of square numbers up to 100 to find the answer. The answer is 9 because $9 \times 9 = 81$.	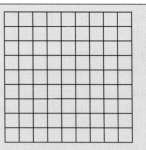 81 is the area of a 9 × 9 square.
d) $\sqrt[3]{64} = 4$	$\sqrt[3]{64}$ means 'the cube root of 64', or 'the product of three copies of what number makes 64?' Use knowledge of cube numbers up to 125 to find the answer. The answer is 4 because $4 \times 4 \times 4 = 64$.	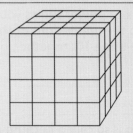 64 is the volume of a 4 × 4 × 4 cube.

Exercise 4 1–3, 5–10

1 Write the value of:

a) 9^2 b) 7^2 c) 5^2 d) 1^2 e) 20^2

f) $\sqrt{4}$ g) $\sqrt{64}$ h) $\sqrt{36}$ i) $\sqrt{100}$ j) $\sqrt{9}$

2 a) Find the area of the square.

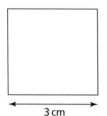

3 cm

b) Find the side length of the square.

Area = 25 cm²

3 Write the value of:

a) 1^3 b) 3^3 c) 4^3 d) 2^3 e) 100^3

f) $\sqrt[3]{8}$ g) $\sqrt[3]{125}$ h) $\sqrt[3]{1000}$ i) $\sqrt[3]{27}$ j) $\sqrt[3]{1}$

4 Use a calculator to find:

a) 25^2 b) 53^3 c) $\sqrt{256}$ d) $\sqrt[3]{1331}$

5 Write these in ascending order.

a) 2^2 5 $\sqrt{9}$ 4^2 $\sqrt{36}$ 12

b) 10^2 50 $\sqrt[3]{125}$ 5^2 $\sqrt{100}$ 10^3

6 Write these in ascending order.

5^3 5 $\sqrt[3]{5}$ 5^2 $\sqrt{5}$

> **Discuss**
>
> Is the square root of an integer larger or smaller than the integer? Is the cube root of an integer larger or smaller than the square root?

7 a) Write a square number that is greater than 70 and less than 90.

b) Write a cube number that is greater than 100 and less than 200.

> **Tip**
>
> In question 7 part **a)**, find the nearest square numbers above and below 6.

8 The square root of 26 lies between 5 and 6, because 26 lies between 5^2 and 6^2.

Between which pair of integers is each of the square roots below?

a) $\sqrt{6}$ b) $\sqrt{30}$ c) $\sqrt{75}$ d) $\sqrt{180}$ e) $\sqrt{112}$

9 a) Nina works out the value of 5.3^2 to be 24.89.
Use estimation to explain why her answer is incorrect.

b) Samir works out the value of 2.8^3 to be 37.848.
Use estimation to explain why his answer is incorrect.

10 Zi Xin is investigating square numbers.

First, he finds the differences between consecutive square numbers:

> **Tip**
>
> 'Consecutive' means in order, with no gaps.

square number	1		4		9		16		25		36
difference		3		5							

a) Copy the lines above and fill in the gaps.

b) Describe the pattern in the differences.

Zi Xin draws diagrams to try to understand the pattern.
The diagrams represent the first six square numbers.

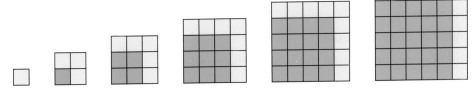

c) Describe how the diagrams show the differences between square numbers.

Thinking and working mathematically activity

Show that 64 is both a square number and a cube number.
Show that 729 and 4096 are also both square and cube numbers.
Can you find a pattern to help identify numbers that are both square numbers and cube numbers?
Use the pattern to find more numbers that are both square numbers and cube numbers.

Consolidation exercise

1 Here are some number cards.

| −7 | | −2 | | 1 | | −4 | | 9 | | 5 |

a) Choose two cards with a sum of 5.

b) Choose two cards with a difference of 2.

c) Choose two cards with a sum of −9.

d) Choose two cards with a product of −18.

2 Which is the odd one out?

a) $-15 - 7$ 　　 b) $-24 - (-2)$ 　　 c) $30 - 52$ 　　 d) $-27 - 5$ 　　 e) $-21 - 1$

3 a) Copy and complete this multiplication grid.

×		−4	
8	40		
		−36	
6			−42

> **Discuss**
>
> Which columns and rows could have different answers? Which could not? Why?

b) Compare your answers with other students.

4 Benji says that $-12 + 35$ has a negative answer, because a negative and a positive make a negative. Explain why Benji is incorrect.

5 Decide whether each statement is always true, sometimes true or never true.

a) If you add an integer to zero, the answer is positive.

b) If you add two negative integers, the answer is negative.

c) If one of the four operations is used with a positive number and a negative number, then the answer is negative.

d) The square of a number does not equal the number.

e) The square of a number does not equal the cube of the number.

6 For each calculation below, one of the four options A–D is the correct answer. Use estimation to choose the correct answers.

a) $223 - (-486)$ 　　 A: −263 　　 B: −163 　　 C: 609 　　 D: 709

b) $23 \times (-81)$ 　　 A: −1863 　　 B: −1603 　　 C: −183 　　 D: −163

c) 625×790 　　 A: 32 750 　　 B: 49 350 　　 C: 323 750 　　 D: 493 750

d) $-517 \div 11$ 　　 A: −47 　　 B: −17 　　 C: 17 　　 D: 47

7 Write the value of:

a) $\sqrt{7^2}$ b) $\sqrt{30} \times \sqrt{30}$

8 Ben's garden is square with an area of $289\,m^2$. He puts a fence around three sides of the garden. What is the total length of the fence?

> **Think about**
>
> Calcuate the values of these expresssions. What do you notice?
>
> a) $1^3 + 2^3$
>
> b) $1^3 + 2^3 + 3^3$
>
> c) $1^3 + 2^3 + 3^3 + 4^3$
>
> Can you predict the value of $1^3 + 2^3 + 3^3 + 4^3 + 5^3$?

End of chapter reflection

You should know that...	You should be able to...	Such as...
Adding a negative number is the same as subtracting a positive number.	Add positive and negative integers.	Find: **a)** $7 + (-6)$ **b)** $-17 + (-12)$
Subtracting a negative number is the same as adding a positive number.	Subtract positive and negative integers.	Find: **a)** $4 - (-2)$ **b)** $-13 - (-9)$
The product of a positive number and a negative number is a negative number.	Multiply positive and negative integers.	Find: **a)** $6 \times (-8)$ **b)** -10×5
A positive number divided by a negative number gives a negative answer.	Divide a positive integer by a negative integer.	Find: $28 \div (-4)$
A negative number divided by a positive number gives a negative answer.	Divide a negative integer by a positive integer.	Find: $-16 \div 8$
You can use rounding to estimate the answers to calculations using the four operations with positive and negative numbers.	Estimate the answers to additions, subtractions, multiplications and divisions with positive and negative numbers.	Estimate: **a)** $-67 + 817$ **b)** $206 \times (-79)$
The square root of a number squares to make the number.	Recognise square numbers to at least 100 and know their square roots.	Write the value of: **a)** 8^2 **b)** $\sqrt{49}$
The cube root of a number cubes to make the number.	Recognise cube numbers to at least 125 and know their cube roots.	Write the value of: **a)** 3^3 **b)** $\sqrt[3]{8}$

5 Expressions

You will learn how to:
- Understand that letters can be used to represent unknown numbers, variables or constants.
- Understand that a situation can be represented either in words or as an algebraic expression, and move between the two representations (linear with integer coefficients).
- Understand that the laws of arithmetic and order of operations apply to algebraic terms and expressions (four operations).

Starting point

Do you remember...

- that the order in which you add two numbers or multiply two numbers does not matter?

  For example, $6 + 5 = 5 + 6$ and $3 \times 4 = 4 \times 3$

- that the order in which you subtract two numbers or divide two numbers does matter?

  For example, $7 - 4$ is not the same as $4 - 7$, and $4 \div 2$ is not the same as $2 \div 4$.

This will also be helpful when...

- you learn how to form and use more complicated algebraic expressions, equations and formulas.

5.0 Getting started

Ask your partner to write down a number between 1 and 20. Tell them not to show you the number.

Give your partner these instructions, but tell them not to speak until you give them these instructions:

'add 3 to your number'

'double your answer'

'add 4 to the answer'

'subtract 10, and tell me your final answer'

Divide their final answer by 2 in your head. Your answer will be their original number.

Try again with a different starting number.

Will this always work? Why?

Can you make up a puzzle of your own to finish with the number that you first thought of?

5.1 Letters for numbers

Key terms

Sometimes you don't know what a number is, but you need a way to represent it. You could leave a space or introduce a symbol for the unknown number. For example, in the activity in Getting started, when you think of a number and want to add three to it, you could write:

... + 3   or $\boxed{}$ + 3

In algebra you can use a letter to represent the unknown number, so you could write:

$m + 3$

$m + 3$ is called an **expression**. m and 3 are both **terms** of the expression.

In this expression m is a **variable**, as its value can change. 3 is a **constant** term, as its value does not change, it will always be 3.

Sometimes there is more than one variable in an expression, for example $a + b$.

You do not know the value of an expression unless you have more information.

An **equation** has an equals sign linking two expressions or numbers that are equal.

$m + 3 = 11$ and $s - 3 = 6$ are both equations. They are true only for a specific value of m and s.

Sometimes you want to multiply a variable by a number. For example, if you want to multiply an unknown number by 4, you could write:

$4 \times$ ....   or $4t$

The 4 is called the **coefficient** of the term, because it's what you multiply your unknown number by.

This means that you can write the expression $4t - 5$ to represent 'multiply a number by 4 then subtract 5' or the equation $3m + 4 = 19$ to represent 'multiply a number by 3 and then add 4 to get an answer of 19'.

Within one equation the same letter will have the same value. However, the same letter can have different values in different equations.

Tip

You do not have to use m, you can use almost any letter.

Discuss

Which letters do you think would not be good to use to represent numbers? Why?

Tip

In algebra, you don't need to write the multiplication sign, you can just write $4t$ instead of $4 \times t$.

Did you know?

Sometimes a letter can be used to represent a constant value. For example, in science, c is used to represent the speed of light.

Worked example 1

Look at the expression $7d + 4$.

a) What does '$7d$' mean?

b) Write the equation that shows that this expression is equal to 18.

a) $7d$ means '$7 \times d$' or 7 lots of d	In algebra you don't need to write the multiplication sign – a number next to a letter always means that you multiply them.	
b) $7d + 4 = 18$	An equation contains an equals sign. The expression is $7d + 4$ It is equal to 18, so $7d + 4 = 18$	

d	d	d	d	d	d	d	4

$7d$	4

$7d$	4

18

Exercise 1

1 **Vocabulary question** Copy and write the correct words in the spaces.

a) $t + 8 = 11$ is an _____.

b) $2r + 6$ is an _____.

c) In $5m$, 5 is called the _____ and m is called the _____.

   $5m$ means _____.

d) In $3d + 8$, $3d$ and 8 are both _____ of the _____.

2 Write down the coefficient of the variable in each of these expressions.

a) $4t - 9$ b) $3 + 8v$ c) $52x + 136$

3 Label the sides of this isosceles triangle with letters to represent the lengths of the side.

4 a) Clive draws a quadrilateral. He labels two sides with the letter a and two sides with the letter b. What could his quadrilateral look like?

b) Dina draws a quadrilateral. She labels two sides with the letter p, one side with the letter q and one side with the letter r. What could her quadrilateral look like?

c) Eric draws a pentagon. He labels three sides with the letter x and two sides with the letter y. What could his pentagon look like?

d) Compare your answers to a), b) and c) with a partner. Is there only one shape possible each time? How many different shapes can you find for each of these questions?

> **Thinking and working mathematically activity**
>
> Explore what different letter combinations you can make for shapes with different numbers of sides.

5.2 Forming expressions

Worked example 2

a) A ribbon of length 7 cm is joined to a ribbon of length n cm and a ribbon of length m cm. Form an expression for the total length of the new ribbon formed.

b) Anna has three lengths of ribbon that are n cm each, plus another length of 7 cm. Find an expression for the total length of ribbon that Anna has.

a) Add 7 and n and m to give the total length of the ribbon. $$7 + n + m$$ or $$m + n + 7$$ Write the answer with the units and a bracket to give $$(7 + n + m) \text{ cm}$$ or $$(m + n + 7) \text{ cm}$$	As you are joining the lengths of ribbon, you add the 7 the n and the m. You can write the terms in any order as the order of addition does not change the answer. **Tip** Writing letters in alphabetical order can help you keep them organised in your answers.	7 | n | m / $7 + m + n$ **Think about** Another answer is $(n + 7 + m)$ cm. How many others can you find?
b) $(3n + 7)$ cm	Three lots of n is $3 \times n$ or $3n$ Add this to 7 to give the total length of ribbon. $$3n + 7$$ Write the answer with units.	n | n | n | 7 / $3n$ | 7

Worked example 3

Pierre, Katie, Otis and Jack are thinking of numbers.

Pierre's number is n.

Write expressions for:

a) Katie's number, if it is seven more than Pierre's number

b) Otis's number, if it is three times Pierre's number

c) Jack's number, if it is one more than Otis's number.

Jack's number is 19.

d) Write an equation to show how Jack's number is linked to Pierre's number.

a) $n + 7$	Pierre's number is n, and Katie's is seven more, so you add seven to n.	n | 7 / $n + 7$
b) $3n$	Pierre's number is n, and Otis's number is three times Pierre's number, which is $3 \times n$, which you write as $3n$.	n | n | n / $3n$

c) $3n + 1$	Otis's number is $3n$ and Jack's is one more than that, so it is $3n + 1$	
d) $3n + 1 = 19$	Jack's number is $3n + 1$, which you is equal to 19.	

Think about

Can you work out what number Pierre thought of?

Exercise 2

1 In each part of this question two parts of a line are shown. Form an expression, in centimetres, for the length of the line that you would get by joining the two parts.

a) _____   _____
   3 cm y cm

b) _____   _____
   t cm   4 cm

c) _____   _____
   7 cm k cm

d) _____   _____
   8 cm x cm

2 Malia is forming expressions.

She writes 'a number 5 more than x' as $x + 5$.

Form an expression from each of these statements.

a) 5 plus y **b)** h minus 8   **c)** the sum of c and 3

d) 6 plus p **e)** 10 minus f **f)** 12 more than t

g) 5 less than k **h)** the difference between 10 and w

3 Form an expression for the perimeter of each of these shapes.

a)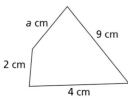

b)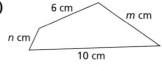

c)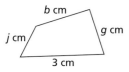

d)

4 Sophie works in a cafe. She works out expressions for the cost of each order.

For example, for two teas and a muffin the cost would be $(2t + m)$ cents.

Write down expressions for the cost of an order of:

a) three coffees and two sandwiches

b) one juice, two sandwiches and a muffin

c) one water, two teas, a pie and two sandwiches.

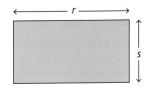

Menu	
Tea, t cents	Pie, p cents
Coffee, c cents	Sandwich, s cents
Juice, j cents	Muffin, m cents
Water, w cents	Biscuit, b cents

5 **a)** The perimeter of this rectangle is 26 cm.
Jane says that this means that $r = 26 - s$.
Is this correct? Explain your answer.

b) Ella says that the perimeter of this shape is $(a + a + b + b + b)$ cm.
Lucas says the perimeter is $(2a + 3b)$ cm. Who is correct?

Explain your answer.

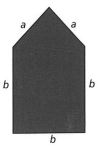

6 **a)** Match each expression with the statement that describes it, then write expressions or equations for the remaining statements. The first one has been done for you.

$\frac{a}{5} + 7$

$8 - \frac{b}{3}$

$5f - 7$

$3 + 8c$

$\frac{d}{3} + 8$

$5p + 7$

$12 - 2e$

Multiply a number by 5 and subtract 7.

Divide a number by 3 then add 8.

Multiply a number by 5 and add 7.

Divide a number by 5 then add 7.

Multiply a number by 4 then add 23.

Divide a number by 3 then subtract your answer from 8.

Multiply a number by 8 and subtract 4.

b) Make three statements of your own and ask a partner to write down the correct expression or equation for each of your statements. Now make three expressions and ask your partner to write the correct statement for each expression. Check your partner's statements – do they match your expressions? Can two different statements ever both be correct for a single expression?

7 Demetrius has x coloured pencils and Cora has y coloured pencils.

a) What might the expression $x + y$ mean?

b) What would the expression $2x$ mean?

c) What would the expression $x - y$ mean? Is this the same as $y - x$?

d) What could $x + 5 = y$ mean?

8 Write a story to match each expression. The first one has been done for you.

a) $3s + \frac{t}{2}$

b) $2m + 1$

c) $d - e$

d) $5t + 4$

a) Eddie has three bags with s apples in each bag and half a box of apples that originally had t apples in it.

Thinking and working mathematically activity

Make a card matching game using expressions written in algebra and in words, similar to those in question 7.

5.3 Order of algebraic operations and substitution

Key terms

There are a number of **operations** in mathematics. You have learned addition, subtraction, multiplication and division.

These operations in algebra work in the same way as they do with ordinary numbers.

For example, $6 + 7 = 7 + 6$ and $a + b = b + a$

Sometimes you want to work out an expression for a particular value of a variable. To do this, you need to **substitute** that value for the letter in the expression.

You can then work out the answer for the expression for that value. Remember to use the correct order of operations in your calculation.

For example, you are told:

- Basil is b years old.
- Aimee is four years older than Basil.
- Zak is twice as old as Basil plus another three years.

You can write algebraic expressions for the ages of Aimee and Zak in terms of b:

- Aimee is $b + 4$ years old.
- Zak is $2b + 3$ years old.

You are now told that Basil is 5 years old.

You can now work out that Aimee is $b + 4 = 5 + 4 = 9$ years old, and Zak is $2b + 3 = 2 \times 5 + 3 = 13$ years old.

Worked example 4

a) Is $7 + p$ the same as $p + 7$?

b) Is $7 - m$ the same as $m - 7$?

c) What is the value of $5t - 7v$ when $t = 4$ and $v = 2$?

d) What is the value of $12 - \frac{w}{3}$ when $w = 21$?

a) Yes, $7 + p$ is the same as $p + 7$ as the order in which you add two numbers does not change the answer.	Think of p as 'any number'. If you add 7 to a number, such as 5, you get the same as when you add the 5 to the 7.	
b) Is $7 - m$ the same as $m - 7$? No.	Think of m as 'any number'. Choose an example, such as $m = 3$. If you do $7 - 3$, you get 4. If you do $3 - 7$, you get -4, which is not the same. You only need to find one example where it is not true for the statement not to be true.	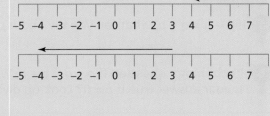
c) $5t - 7v$ $= 5 \times 4 - 7 \times 2$ $= 20 - 14$ $= 6$	You need to work out $5 \times t$ and $7 \times v$ separately. You then subtract the answer to $7 \times v$ from the answer to $5 \times t$. Remember to use the correct order of operations. You are told that t is 4 and v is 2.	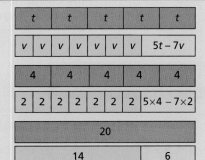
d) $12 - \frac{w}{3}$ $= 12 - \frac{21}{3}$ $= 12 - 7$ $= 5$	$\frac{w}{3}$ means 'w divided by 3'. You need to work out $\frac{w}{3}$ first then work out 12 minus the answer to your division. Remember to use the correct order of operations. You are told w is 21.	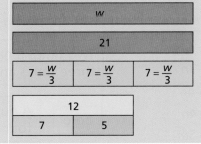

Exercise 3

1 Which of the following are true and which are false?

a) $5 + 6 = 6 + 5$ **b)** $6 - 5 = 5 - 6$ **c)** $6 - 4 + 3 = 4 + 6 - 3$

d) $6 \div 2 = 2 \div 6$ **e)** $3 \times 6 \div 2 = 2 \div 6 \times 3$ **f)** $3 \times 6 \div 2 = 6 \times 3 \div 2$

g) $6 \times n = n \times 6$ **h)** $3 \times q \times 4 = 4 \times 3 \times q$ **i)** $2 \times h \div 4 = 4 \times 2 \div h$

2 If $p = 2$ and $q = 6$, find the value of:

a) $3 + 2p$ b) $20 - 3q$ c) $p + q$ d) $2p + q$

e) $3q - 6p$ f) $\frac{q}{2}$ g) $p - \frac{q}{3}$ h) $4p + 6q - 20$

3 If $t = 30$, $u = 50$ and $v = 60$, calculate:

a) $v + 2u$ b) $8 + \frac{t}{5}$ c) $\frac{v}{10} - \frac{t}{6}$ d) $v + 3t - 2u$

4 If $d = 2$, $e = 8$ and $f = 12$, sort the expressions into groups of the same value.

a) $2d + f$ b) $\frac{f}{2} - 3d$ c) $2d + 1$ d) de

e) $\frac{e}{4} + 3$ f) $2e$ g) $2f - 3e$ h) $2e - f + 1$

i) $3d - 1$ j) $e - 4d$ k) $3e - 4d$

Make up two more expressions for each group using d, e, and f with the values given above.

5 If $p - q = q - p$, what can you say about p and q? Is your statement always true?

6 **Technology question** If $p \div q = q \div p$, what can you say about the values of p and q? Is your answer true if $p = 0$? Look up division by zero on the internet.

Thinking and working mathematically activity

Using the values $a = 2$, $b = 3$, $c = 4$ and $d = 5$, how many different expressions can you make that will give a total of 24? Think about how you can check that the expressions are different, not just written in a different order (for example, abc would be considered the same as bac).

Consolidation exercise

1 Are the following statements true or false for all possible values of the variables?

a) $4c$ means $4 \times c$ b) $c - d = d - c$ c) $a + b = b + a$

d) $\frac{m}{3}$ means 'm divided by 3'.   e) $3 \times d = d \times 3$ f) $10 \div h = h \div 10$

2 Match these expressions with the phrase that describes them.

a) $3m + 4$

b) $2m - 3n$

c) $2m + 4n$

d) $\frac{m}{3} + 4n$

e) $5m - 5m$

| Two times a number minus three times another number | Two times a number plus four times another number |

Zero

| A number divided by three plus four times another number | Three times a number plus four |

3 Write expressions for the perimeter of these shapes.

a)

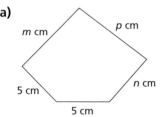

b)

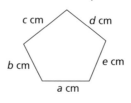

4 Find the perimeter of the shape in part **a)** of question 3 if:

a) $m = 5$, $n = 4$ and $p = 6$ b) $m = 6$, $n = 5$ and $p = 8$

5 If $e = 5$, $f = 7$ and $g = 4$, write expressions that can give an answer of:

a) 16   b) 33   c) 2   d) 0   e) 50

Try to write at least three different expressions for each answer.

End of chapter reflection

You should know that...	You should be able to...	Such as...
You can use letters to represent unknown numbers, variables or constants.	Identify the terms in an algebraic expression. Write down the coefficient of a variable.	How many terms are there in the expression $4t + 5$? Write down the coefficient of t in $4t + 5$.
Algebraic expressions can be used to represent real situations.	Construct algebraic expressions.	Anya is a years old. Maria is twice as old as Anya. Write an expression in terms of a for Maria's age.
The laws of arithmetic and order of operations apply to algebraic terms and expressions.	Substitute numbers into expressions.	Work out $s + 6t$ when $s = 4$ and $t = 3$. In $e + 3f$ you calculate the $3 \times f$ before you do the addition.

6 Symmetry

You will learn how to:
- Identify, describe and sketch regular polygons, including reference to sides, angles and symmetrical properties.
- Identify reflective symmetry and order of rotational symmetry of 2D shapes and patterns.

Starting point

Do you remember…

- the names of polygons with 3, 4, 5, 6 and 8 sides?

  For example, a shape with 6 sides is called a hexagon.
- what a regular polygon is?

  For example, a regular pentagon is a five-sided shape with equal sides and angles.
- how to find a line of symmetry?

  For example, draw the lines of symmetry on this shape.
- what rotational symmetry is?

  For example, find the order of rotational symmetry for this same shape.

This will also be helpful when…

- you learn more about the symmetries of regular polygons.

6.0 Getting started

Here are six triangle pieces.

Some or all of these pieces can be put together to make geometric shapes, such as those shown. Can you name these three shapes?

How many different geometric shapes can you make using any or all of the six pieces?

Draw these shapes and name them.

You could use isometric paper to draw your shapes.

Name two shapes that you cannot make from these pieces.

Explain your answer.

6.1 Polygons

Name of regular polygon (and number of sides)	Diagram	Number of lines of symmetry	Order of rotational symmetry
Equilateral triangle (3 sides)		3	3
Square (4 sides)		4	4
Regular pentagon (5 sides)		5	5
Regular hexagon (6 sides)		6	6
Regular heptagon (7 sides)		7	7
Regular octagon (8 sides)		8	8
Regular nonagon (9 sides)		9	9
Regular decagon (10 sides)		10	10

Key terms

A **polygon** is a closed, two-dimensional shape made of straight sides. A polygon is said to be **regular** if all of its sides have equal length and all of its angles are equal.

Think about

Will every regular polygon have the same number of lines of symmetry as its order of rotational symmetry?

How many lines of symmetry are in a regular polygon that has rotational symmetry of order 100?

Discuss

Describe the symmetries of a circle.

Worked example 1

$ABCDEF$ is a regular hexagon.

a) Sketch the regular hexagon $ABCDEF$.

b) Describe the pairs of parallel sides of $ABCDEF$.

c) What type of quadrilateral is $ABEF$? Explain your answer.

d) How many lines of symmetry does $ABCDEF$ have?

e) What is the order of rotational symmetry for $ABCDEF$?

Tip

To 'sketch' means that you do not have to measure the angles and sides exactly but they should look reasonable.

a)

A regular hexagon has all its six sides and angles equal.

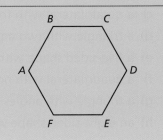

b) *ABCDEF* has three pairs of parallel sides: *AB* is parallel to *ED* *BC* is parallel to *FE* *CD* is parallel to *AF*	Regular polygons with an even number of sides have opposite sides parallel.	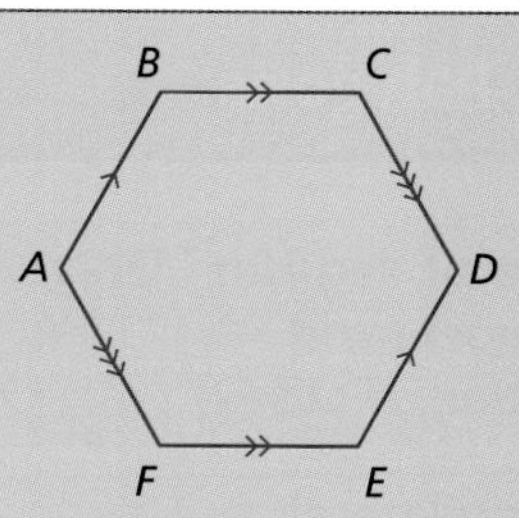
c) *ABEF* is a trapezium as it is a quadrilateral with one pair of parallel sides.	*ABEF* has four sides so it is a quadrilateral. *AF* is parallel to *BE* but *AB* is not parallel to *FE* so *ABEF* has one pair of parallel sides. A quadrilateral with one pair of parallel sides is a trapezium.	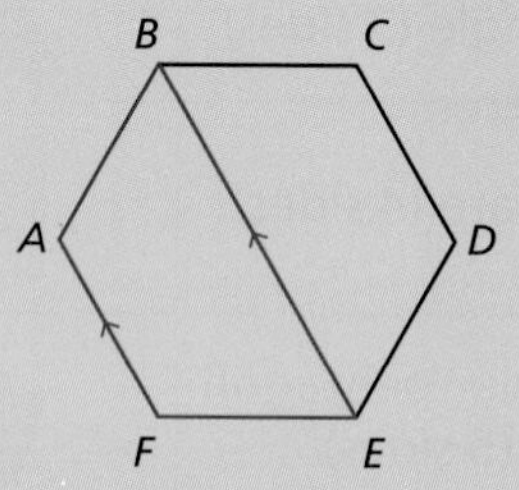
d) A regular hexagon has six lines of symmetry.	There are six different ways that the hexagon can be folded in half.	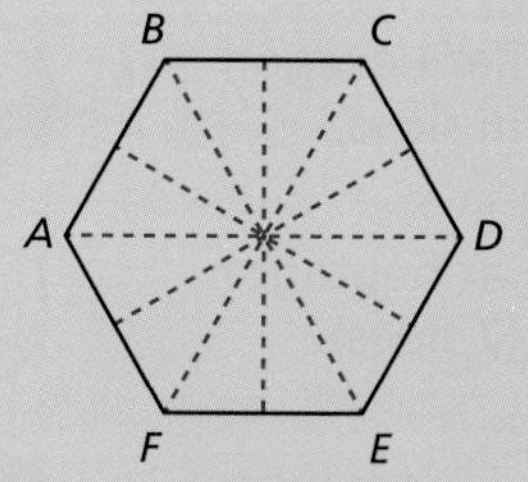
e) A regular hexagon has rotational symmetry of order six.	The hexagon can be rotated around its centre six times and each time it looks the same as the original position. One such rotation is shown here.	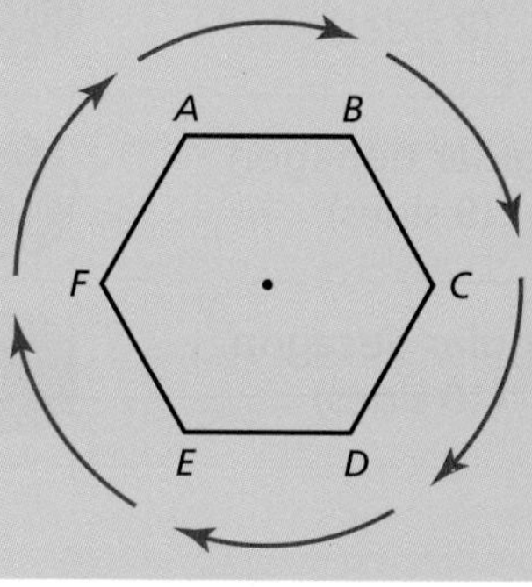

Exercise 1

1 Name these shapes from their descriptions.

  a) a triangle with all its sides the same length

  b) a quadrilateral with four equal sides and four right-angles

  c) a quadrilateral with four equal sides and no right-angles

  d) a triangle with two equal angles

  e) a five-sided shape with all of its angles equal

  f) a quadrilateral with no parallel sides but one pair of equal angles

  g) a triangle with angles of 40°, 50° and 90°

  h) an eight-sided shape with all sides of equal length.

2 Sort these shapes into the correct place on the two-way table.

rectangle   equilateral triangle   regular hexagon

kite   isosceles triangle   rhombus

trapezium   scalene triangle   square

	Quadrilateral	Not a quadrilateral
Regular		
Irregular		

Thinking and working mathematically activity

The regular hexagon shown has a single blue line drawn on it which divides the hexagon into two quadrilaterals.

- On a copy of this hexagon, draw a single straight line to make an isosceles triangle and a pentagon.

- Can a single straight line be drawn on a regular hexagon to make an isosceles triangle and a hexagon?

- On a copy of this hexagon, draw a single straight line to make two pentagons.

- What other geometric shapes can you make by drawing a single straight line on a regular hexagon?

- What shapes can you make by drawing a single straight line on a regular octagon?

3 Copy and complete each of these statements.

a) A square has _____ sides.

_____ of these sides are equal in length and it has _____ pairs of parallel sides.

b) An equilateral triangle has _____ sides.

_____ of these sides are equal in length and it has _____ pairs of parallel sides.

c) A regular pentagon has _____ sides.
_____ of these sides are equal in length and it has _____ pairs of parallel sides.

d) A regular decagon has _____ sides.
_____ of these sides are equal in length and it has _____ pairs of parallel sides.

e) A regular nonagon has _____ sides.
_____ of these sides are equal in length and it has _____ pairs of parallel sides.

4 Match the polygon to a correct statement about its symmetries. Complete the missing shape and symmetry.

Polygon
square
regular pentagon
regular octagon
regular nonagon
regular decagon

Symmetry
nine lines of symmetry
rotational symmetry of order 4
five lines of symmetry
rotational symmetry of order 7
eight lines of symmetry

5 Sketch and name each of these regular polygons.

    a) a regular polygon that has seven angles of equal size

    b) a regular polygon that has six sides of equal length

    c) a regular polygon that has five lines of symmetry

    d) a regular polygon with 9 lines of symmetry and rotational symmetry of order 9.

6 Anahita draws a regular pentagon. Olly draws in all the lines of symmetry. How many triangles has the shape now been divided into?

7 Andrew and Sophia have been asked to sketch an octagon. Andrew sketches a regular octagon. Sophia sketches an irregular octagon with rotational symmetry of order 2.

    a) Draw sketches of the octagons drawn by Andrew and Sophia.

    b) Describe how their octagons are different.

8 **a)** Sketch a hexagon with 2 lines of symmetry.

    b) Sketch a decagon with rotational symmetry of order 2.

6.2 Line and rotation symmetry

Worked example 2

For each shape, write down:

i) the number of lines of symmetry it has

ii) the order of rotational symmetry.

 a)   b)    c)

a) i) It has five lines of symmetry as it can be folded in half along the five lines shown. **ii)** It has rotational symmetry of order five as it will fit back onto its outline five times in a full turn.	
b) i) It has two lines of symmetry as it can be folded in half along a vertical line and a horizontal line. **ii)** It has rotational symmetry of order two as it will fit back onto its outline twice in a full turn.	
c) i) It has one line of symmetry as it can only be folded in half along a vertical line. **ii)** It has rotational symmetry of order 1 as it will only fit back onto its outline after a full turn.	**Note:** Order 1 rotational symmetry is an equivalent way of saying the shape has no rotational symmetry.

Exercise 2

1 Name two quadrilaterals with exactly two lines of symmetry.

2 For each of these shapes, state how many lines of symmetry it has and its order of rotational symmetry.

a)
b)
c)
d)

e)
f)
g)
h)

3 For each shape:

   a) state the number of lines of symmetry

   b) state the order of rotational symmetry.

i)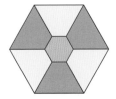
ii)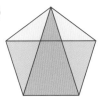
iii)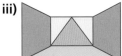
iv)

v)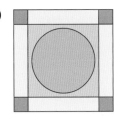
vi)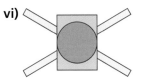
vii)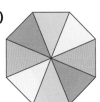
viii)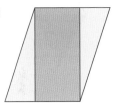

4 For each of these patterns, state how many lines of symmetry it has and its order of rotational symmetry.

a)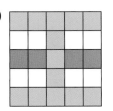
b)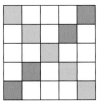
c)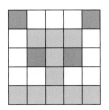
d)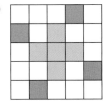

e)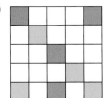
f)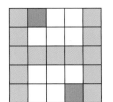
g)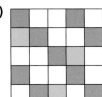
h)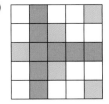

5 Copy and complete each diagram so that the finished pattern has rotational symmetry of order 4. The centre is marked.

a)

b)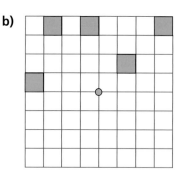

c)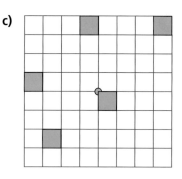

6 a) Draw two quadrilaterals, both with two lines of symmetry and rotational symmetry of order 2.

b) Salim says there is no quadrilateral with one line of symmetry and rotational symmetry of order 1.

Is Salim correct? Give a reason for your answer.

7 Copy this shape on isometric paper and shade four more triangles so that the shape has rotational symmetry of order 3.

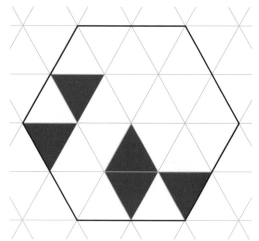

▼ Thinking and working mathematically activity

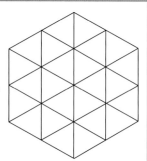

- Shade in six triangles to make a pattern with rotational symmetry of order 3 and three lines of symmetry.

- Try to produce patterns with different symmetrical properties. Each time you should shade in six triangles. Describe the symmetries of the shapes you have drawn.

8 Jane says that a parallelogram has two lines of symmetry.

Do you agree? Explain you answer.

9 Create your own shape which has rotational symmetry of order 3.

Try to make your design intricate and colourful.

10 Describe the symmetries of each of these shapes.

a)

b)
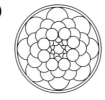

c)

d)

e)

f)

11 Make a copy of a 3 by 3 grid, like this one.

a) Can you shade in one square to give rotational symmetry? How many ways can this be done?

b) Can you shade in two squares to give rotational symmetry? How many ways can this be done?

c) Laz said, 'It does not matter how many squares you want to shade, you can always give it rotational symmetry.' Is Laz correct? Explain your answer.

> **Did you know?**
>
> Every snowflake is different and they all have six lines of symmetry and rotational symmetry of order six.

Consolidation exercise

1 State the symmetries of the following shapes:

a) square   b) parallelogram   c) regular octagon   d) regular decagon

2 For each of these shapes, state how many lines of symmetry it has and its order of rotational symmetry.

a)

b)

c)

d)

e)

3 **a)** Draw four shapes, each with exactly two lines of symmetry.

b) What rotational symmetry does each shape have?

c) Do you think that every shape with two lines of symmetry has rotational symmetry of order 2?

d) Does every shape with rotational symmetry of order 2 have two lines of symmetry? Explain your answer.

4 Which of these can you draw?

a) a trapezium with a line of symmetry

b) a trapezium with rotational symmetry

c) a shape with no lines of symmetry but having rotational symmetry of order five.

5 Put these shapes in order of how many lines of symmetry they have, starting with the parallelogram.

regular octagon   square   parallelogram

circle   equilateral triangle   rhombus

6 **Technology question** Alloy wheels for cars are often made in patterns which have line or rotational symmetry. Here are some pictures of alloy wheels.

Ignoring the wheel nuts, find the number of lines of symmetry and order of rotational symmetry for each wheel. Use the internet to find other pictures of alloy wheels and find as many different orders of rotational symmetry as you can.

a)

b)
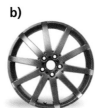

c)

End of chapter reflection

You should know that...	You should be able to...	Such as...
All the sides in a regular polygon have an equal length and all the angles are the same size.	Remember the names of regular polygons and write down their properties. Sketch regular polygons.	Write down the name of the polygon which has six equal length sides and six equal angles.
An object has line symmetry if it can be folded in half exactly along a straight line. The order of rotational symmetry is the number of times a shape or patterns fits back onto into itself when rotated through one full turn.	Describe the symmetry properties of patterns and polygons.	Describe the symmetry properties of this shape.

7 Rounding and decimals

You will learn how to:

- Use knowledge of place value to multiply and divide whole numbers and decimals by any positive power of 10.
- Round numbers to a given number of decimal places.
- Estimate, add and subtract positive and negative numbers with the same or different number of decimal places.
- Estimate, multiply and divide decimals by whole numbers.

Starting point

Do you remember…

- how to multiply and divide positive integers by 10, 100 or 1000 using knowledge of place value?
  For example, work out $450 \div 1000$
- how to round numbers with two decimal places to the nearest tenth or whole number?
  For example, round 3.77 to the nearest tenth.
- how to add and subtract positive numbers with the same or different numbers of decimal places?
  For example, work out $3.8 + 1.45$
- how to multiply positive decimals by positive 1-digit and 2-digit integers?
  For example, work out 8.4×13
- how to divide positive numbers with one or two decimal places by positive integers?
  For example, work out $2.64 \div 12$
- how to estimate results of integer calculations that use one of the four operations?
  For example, estimate $388 \times (-41)$

This will also be helpful when…

- you learn to multiply and divide numbers by 0.1 and 0.01
- you learn to multiply decimals by decimals
- you learn to divide decimals by numbers with one decimal place.

7.0 Getting started

Use the digits 1, 2, 3 and 4 once each to complete the addition.

- What is the highest result you can make? What is the lowest?
- What is the closest result to 5?
- How many different results are there?
- What happens to the answer if you swap the tenths digits?

Answer the same questions for the subtraction below – but only consider subtractions with positive results.

Use the digits 1, 2, 3 and 4 once each to complete the multiplication.

- What is the highest result you can make? What is the lowest?
- What is the closest result to 5?
- Does the result change if you swap the units digits?

Key terms

A **power** or **index** tells you how many copies of a number to multiply together. For example, 10^4 means $10 \times 10 \times 10 \times 10 = 10\,000$. 10^4 is called 'the fourth power of 10' or '10 to the power of 4' (or '10 to the 4' for short). The 4 is the index or power of 10.

7.1 Multiplying and dividing by positive powers of 10

Tip

Note that $10^1 = 10$

Worked example 1

a) Write the value of 10^0

b) Write the value of 10^4

c) Write 1000 as a power of 10

d) Find 36×10^4

e) Find 0.036×10^4

f) Find $0.2 \div 10^4$

Did you know?

The number 10^{100} is called a googol. The internet search engine Google was named after this huge number, because Google aims to provide very large amounts of information.

a) $10^0 = 1$	'10 to the power of 0' equals 1.	$\times 10$ $10^3 = 1000$ $\div 10$ $\times 10$ $10^2 = 100$ $\div 10$ $\times 10$ $10^1 = 10$ $\div 10$ $10^0 = 1$
b) $10^4 = 10 \times 10 \times 10 \times 10$ $= 10\,000$	'10 to the power of 4' means multiply four 10s together.	10^4 branches to $10_1 \times 10_2 \times 10_3 \times 10_4 = 10\,000$ (1 2 3 4)
c) $10 \times 10 \times 10 = 10^3$	The product of three 10s can be written as '10 to the power of 3' (or '10 cubed').	1000 (1 2 3) branches to $10_1 \times 10_2 \times 10_3 = 10^3$
d) $36 \times 10^4 = 36 \times 10\,000$ $= 360\,000$	Multiplying by 10^4 is the same as multiplying by 10 four times, or multiplying by 10 000. Each digit moves to the left four times. Write zeros in the empty spaces up to the decimal point.	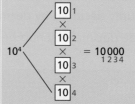

e) $0.036 \times 10^4 = 0.036 \times 10\,000$ $= 360$		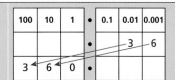
f) $0.2 \div 10^4 = 0.2 \div 10\,000$ $= 0.00002$	Dividing by 10^4 is the same as dividing by 10 four times, or dividing by 10 000. Each digit moves to the right four times. Write zeros in the empty spaces.	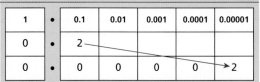

Exercise 1 1–7

1 Write as ordinary numbers:

a) 10^3 b) 10^6 c) 10^7 d) 10^0

2 Write as powers of 10:

a) 10×10 b) $10 \times 10 \times 10 \times 10 \times 10$ c) $10 \times 10 \times 10 \times 10$ d) $100\,000$

e) ten thousand   f) 1000 g) 1 h) 10

3 Write these numbers in order of size, smallest first.

10^4 $1\,000\,000$ 10^5 1000

4 Find:

a) 4×10^2 b) 5.0024×1000 c) 56×10^4 d) 0.25×10^5

e) 3.2×10^3 f) 0.0098×10^6 g) 6.8223×10^7 h) 72×10^0

i) $81 \div 10^2$ j) $0.14 \div 10^3$ k) $3650 \div 10^4$ l) $680\,000 \div 10^5$

m) $82.5 \div 10^3$ n) $980 \div 10^6$ o) $68.2 \div 10^7$ p) $1 \div 10\,000$

5 Match the pairs of expressions with equal value.

A 4×10^5 B $4 \div 10^4$ C 0.0004×10^5 D 0.0004×10^4

E $400\,000 \div 10^4$ F 0.00004×10^1 G $400 \div 10^2$ H 400×10^3

6 Emily is working out this calculation: 4.314×10^2. She says, 'I just need to add two zeroes on to the end of my number, so the answer is 4.31400.'

Do you agree with Emily? Explain your answer.

7 There are mistakes in the statements below.

i) $8 \times 10^3 = 0.8 \times 10^2$ ii) $2300 \div 10^3 = 2.3 \times 10$ iii) $10^1 \times 10^0 = 1$

a) Describe the mistakes and correct them.

b) Is there only one possible way to correct them? Explain your answer.

8 Find the index or power key on your calculator. (It may have the symbol ^ or x^y.) This button is used to input powers, such as 10^6. Use this button on your calculator to do the calculations below.

a) 10^4　　　　　　b) 5.621×10^6　　　　　c) $403 \div 10^3$　　　　　d) 0.0045×10^5

9 Match the pairs of equal numbers.

10^6
Ten to the power of 1
10^5
Ten to the power of four

Ten
One hundred thousand
Ten Thousand
One million

Thinking and working mathematically activity

These calculations involve multiplying by numbers that are close to powers of ten:

102×55　　　6×99　　　98×99　　　444×1001　　　82×999　　　25×9998

Look for quick ways to do these multiplications. Explain your methods, and show that they are quicker than using long multiplication.

7.2 Rounding to decimal places

Key terms

A **decimal place** is a space after the decimal point in a number where you can write a digit. The number 3.142 has three decimal places (3 d.p.) because it has three digits after the decimal point.

Accurate means close to the true value. Rounding a number usually makes it less accurate. The fewer places you round to, the less accurate the rounded value is.

Tip

Rounding to one decimal place (1 d.p.) is the same as rounding to the nearest tenth.

Rounding to two decimal places (2 d.p.) is the same as rounding to the nearest hundredth.

Worked example 2

Round the number 9.8257:

a) to one decimal place (1 d.p.)

b) to three decimal places (3 d.p.)

a) 9.8257 to one decimal place is 9.8	Rounding to one decimal place is the same as rounding to the nearest tenth. 9.8257 lies between 9.8 and 9.9. The digit in the second decimal place is 2, so don't round the 8 up.	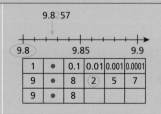

b) 9.8257 to three decimal places is 9.826	Rounding to three decimal places is the same as rounding to the nearest thousandth. 9.8257 lies between 9.825 and 9.826. The digit in the fourth decimal place is 7, so round the 5 up to 6.	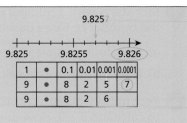

Exercise 2 — 1–4 and 7

Discuss

Not everyone agrees on how to round negative numbers. For example, some people round -1.5 down to -2, and other people round it up to -1. Think of reasons for using each of these methods.

1 Round these numbers to one decimal place.

   a) 0.46   **b)** 6.413   **c)** 12.389

   d) 21.01   **e)** 7.98

2 Round each number to the accuracy shown.

   a) 0.386 (2 d.p.)   **b)** 72.6088 (3 d.p.)   **c)** 1.50923 (4 d.p.)

   d) 5.55555 (4 d.p.)   **e)** 0.00014 (4 d.p.)   **f)** 670.4009 (3 d.p.)

3 Here is a number: 0.068506

   a) Which of these statements are correct?

A:	**B:**	**C:**	**D:**	**E:**
The number rounded to 1 d.p. is 0.0	The number rounded to 2 d.p. is 0.07	The number rounded to 3 d.p. is 0.068	The number rounded to 4 d.p. is 0.0685	The number rounded to 5 d.p. is 0.06851

   b) Correct the statements that are not correct.

4 Gautam and Amelia try to round the number 6.999 to two decimal places.

Gautam says, 'There is a 9 in the third decimal place, so round the 9 in the second decimal place up to 10. The answer is 6.910.'

Amelia says, 'You cannot round a digit up to 10, so the answer is 6.99.'

Gautam and Amelia are both incorrect. Explain how to round 6.999 to two decimal places, and write the correct answer.

5 Use a calculator to do these calculations. Round each answer to three decimal places.

   a) $1 \div 3$ **b)** $5 \div 16$ **c)** $8 \div 9$ **d)** $15 \div 19$

   e) $\sqrt{6}$ **f)** 4.56^2 **g)** $\sqrt{19}$ **h)** $\sqrt[3]{5}$

6 Use a calculator to solve these problems.

   a) Seven people share $123.00 equally. Work out how much each person gets, to the nearest cent. (A cent is one hundredth of a dollar.)

   b) A rectangular field is 12.82 metres wide and 30.14 metres long. Work out the area of the field in square metres, rounded to three decimal places.

7 Jack starts with a number with four decimal places. He rounds the number to three decimal places. His answer is 7.862.

a) Give four possible values for the number Jack started with.

b) How many possible values are there for Jack's number?

Tip

If you are not sure, compare your answer to part **a)** with other students' answers.

8 **Technology question** Use a spreadsheet program, such as Excel, to calculate $8 \div 9$. How does the program round the answer? Compare with the answer on a calculator.

Find out how to make the spreadsheet round the answer to a different number of decimal places.

Thinking and working mathematically activity

Helen says that 0.548 rounded to one decimal place is 0.6. This is her working:

0.548 rounded to two decimal places is 0.55

0.55 rounded to one decimal place is 0.6

Helen's method is not the method you have learned. Her method is called 'double rounding'.

Investigate double rounding, and look for examples when double rounding agrees with your method of rounding and examples of when it disagrees. Explain your findings.

Many people say that double rounding is an incorrect rounding method – why do you think this is?

7.3 Adding and subtracting decimals

Did you know?

In some countries the decimal point is written as a full stop, but in other countries a comma is used instead.

Worked example 3

Find:

a) $-7.3 + 0.5$

b) $3.2 - 5.9$

a) $-7.3 + 0.5 = -6.8$	Start at -7.3 on the number line. Move 0.5 to the right. You can do this in one or more steps.	number line showing +0.3 then +0.2 from −7.3 to −7.0 to −6.8
b) $3.2 - 5.9 = -2.7$	Start at 3.2 on the number line. Move 5.9 to the left. Alternatively, notice that $3.2 - 5.9$ is the negative of $5.9 - 3.2$. Use the column method to calculate this, and then write a negative sign with the answer.	number line showing −0.7, −0.2, −2, −3 moving left from 3.2, 0.2, −1.8, −2, −2.7

Worconst example 4

Estimate, and then find:

a) −13.8 − 7.56

b) 36.18 − 45.62

a) $-13.8 - 7.56 \approx -14 - 8$ $\qquad\qquad\qquad = -22$	Estimate the answer. −13.8 is closer to −14 than to −13. −7.56 is closer to −8 than to −7.	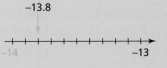 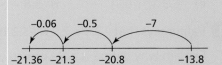
$-13.8 - 7.56$ $= -13.8 - 7 - 0.5 - 0.06$ $= -21.36$	You could use a number line as shown on the right.	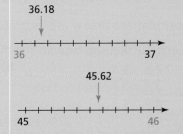 (This diagram is not to scale.)
Alternative method: $-13.8 - 7.56 = -(13.8 + 7.56)$ $\overset{1\ \ 1}{1\ 3} . 8\ 0$ $\underline{+\ \ \ 7 . 5\ 6}$ $\ \ 2\ 1 . 3\ 6$ $13.8 + 7.56 = 21.36$ So $-13.8 - 7.56 = -21.36$	Alternatively, you could use the column method. −13.8 − 7.56 is the negative of 13.8 + 7.56. So work out 13.8 + 7.56 and then write a negative sign on the result.	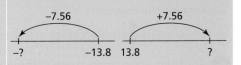
b) $36.18 - 45.62 \approx 36 - 46$ $\qquad\qquad\qquad\quad = -10$	Estimate the answer. 36.18 is closer to 36 than to 37. 45.62 is closer to 46 than to 45.	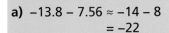
$36.18 - 45.62 = -(45.62 - 36.18)$ $\overset{3\ \ 1\ \ \ 5\ \ 1}{\cancel{4}\ 5 . \cancel{6}\ 2}$ $\underline{-\ 3\ 6 . 1\ 8}$ $\ \ \ 9 . 4\ 4$ $45.62 - 36.18 = 9.44$ So $36.18 - 45.62 = -9.44$	The answer will be negative, so you cannot write 36.18 − 45.62 as a column subtraction. But 36.18 − 45.62 is the negative of 45.62 − 36.18. So calculate 45.62 − 36.18 and then write a negative sign on the result.	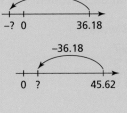

Exercise 3 | 1–6 and 8–11

1 Find:

a) −0.5 + 3.5   b) 0.2 − 0.7   c) −0.3 − 0.5   d) 0.6 + (−1)

2 Draw a number line to show the numbers from −1 to +1 in steps of 0.1.
Mark the answers to each problem on your number line.

a) −2.3 + 2.5   b) 0.4 − 0.7   c) −0.9 + 1.3   d) 0.3 − (−0.4)

e) −0.6 − (−0.2)   f) −0.5 − (+0.3) − 0.6   g) −0.7 − (−0.2) + 0.4   h) −0.7 − (−0.2) − 0.4

3 Find the answers to these. Use a number line if you need to.

a) 8.3 + 2.4   b) −2.7 − 1.5   c) −4.1 + 6.1   d) 5.6 − (3.9 + 7.9)

e) −6.3 − (+ 6.7)   f) 5.8 − (−6.5 + 2.1)   g) 1.1 + (−10.5)   h) −6.7 − (−3.2)

4 Valma says that −3.1 + 2.2 = −5.3 because a positive and a negative make a negative.

Explain why Valma is not correct.

5 The difference between two numbers is 4.7

One of the numbers is 3.9

How many possible values are there for the other number? Find the values.

6 Find the answers to these using a number line or a written method.

a) 3.26 + 5.1   b) −6.58 − 3.3   c) −8.4 + 9.47   d) 7.8 − (2.2 + 8.65)

e) −12.76 − (+6.6)   f) −5.28 − (−5.28)   g) 3.52 + (−2.4)   h) −8.07 − (−0.9)

7 Use a calculator to find:

a) 4.56 − 8.68   b) −3.26 − (+7.851)   c) −5 − (−2.4007)   d) −28.93 − (1.5 − 1.4618)

8 The calculations below are incorrect.

For each calculation, explain the mistake and find the correct answer.

a)
```
  2 7 5 . 9̶⁸ ⁴¹
−   3 4 . 5 7
─────────────
    4 9 . 3 7
```

b)
```
    2 0 4 . 6 5
+     3 9 . 3 7
───────────────
    2 4 3 1 0 2
```

9 Habiba, James and Kiyoko want to calculate 6.57 − 9.23. Their workings are shown below.

Describe what each student has done incorrectly.

a) Habiba's working:	b) James's working:	c) Kiyoko's working:
$\begin{array}{r} 6.57 \\ +\ 9.23 \\ \hline 15.80 \end{array}$ 6.57 − 9.23 = −(6.57 + 9.23) = −15.8	$\begin{array}{r} 6.57 \\ -\ 9.23 \\ \hline -\ 3.34 \end{array}$ 6.57 − 9.23 = −3.34	$\begin{array}{r} ⁸9.²¹²3 \\ -\ 6.57 \\ \hline 2.66 \end{array}$ 6.57 − 9.23 = 9.23 − 6.57 = 2.66

d) Write the correct answer to 6.57 − 9.23.

10 For each calculation, estimate the answer by rounding each number to the nearest integer or nearest ten. Then find the exact answer using a written method.

a) $4.52 + 3.96$ b) $-2.87 - 7.11$ c) $-3.72 + 8.284$ d) $1.445 - 4.073$

e) $-23.6 - (-5.43)$ f) $13.79 + (-8.88)$ g) $-49.84 - (-52.36)$ h) $-33.093 - (+19.21)$

11 For each problem, first estimate the answer. Then use any method (without a calculator) to find the exact answer.

a) The temperature in a room was 22.3 °C. A window was opened and the temperature decreased by 3.5 °C. The window was then closed and the temperature increased by 4.1 °C.

Find the new temperature in the room.

b) On one day in December, the maximum temperature in Adelaide was 28.2 °C and the maximum temperature in Ulaanbaatar was −19.4 °C.

Find the difference between these two temperatures.

c) Heights below sea level can be written as negative numbers. Heights above sea level can be written as positive numbers. Steven travels from Lowtown at height −8.9 m to Highville, which is 127.3 m higher.

Find the height above sea level of Highville.

d) Mishal takes 11.36 seconds to run a race.

Keira takes 0.4 seconds less than Mishal.

Seema takes 1.09 seconds more than Keira.

Find the sum of the times taken by the the three runners.

> **Think about**
>
> Look for quick ways of doing these calculations in your head:
>
> $7.68 + 0.99$
>
> $-7.68 + 0.99$
>
> $5.612 - 4.999$
>
> $-5.612 - 4.999$

Thinking and working mathematically activity

In the calculations below, the decimal point has been missed out of some of the numbers.

Write decimal points to make each calculation correct. (There is more than one possible way to do this for each calculation.)

a) $4\ 5 + 6\ 2 = 6\ 2\ 4\ 5$ b) $6\ 1 + 3\ 7 = 6\ 4\ 7$ c) $1\ 2\ 3 + 2\ 3\ 4 = 1\ 4\ 6\ 4$

d) $5\ 7 - 6\ 3\ 2 = -\ 6\ 2$ e) $2\ 6\ 7 - 5\ 3\ 9 = -5\ 1\ 2\ 3$ f) $6\ 8\ 3\ 4 - 2\ 5\ 2 = 6\ 8\ 0\ 8\ 8$

Discuss the strategies you used. Create your own puzzles like these.

7.4 Multiplying and dividing decimals

Key terms

A **dividend** is a number you are dividing. The **divisor** is the number you are dividing by. The answer to a division calculation is called the **quotient**. For example, in the calculation $56 \div 10 = 5$ with **remainder** 6, the dividend is 56, the divisor is 10, and the quotient is 5.

Short division is a written division method where you work out remainders in your head.

Long division is a written division method where you work out remainders on paper.

Worked example 5

Estimate and then calculate 41.23 × 207

41.23 × 207 ≈ 40 × 200 = 8000	First, estimate the answer. 41.23 lies between 40 and 50, and it is closer to 40. 207 lies between 200 and 300, and it is closer to 200.	41.23 ├─────────────────┤ 40 50 207 ├─────────────────┤ 200 300
$\begin{array}{r} 4\ 1\ 2\ 3 \\ \times \quad\ 2\ 0\ 7 \\ \hline 2\ 8\ 8\ 6\ 1 \\ 8\ 2\ 4\ 6\ 0\ 0 \\ \hline 8\ 5\ 3\ 4\ 6\ 1 \end{array}$	When you multiply a decimal, first ignore the decimal point and write an integer multiplication.	To show the multiplication of each part of 4123 (4000, 100, 20 and 3) by each part of 207 (200 and 7), you can use a grid: |	4000	100	20	3		
---	---	---	---	---	---	---
7	28000	700	140	21	=	28861

200	800000	20000	4000	600	=	824600	
							
							                     +
							                 853461

41.23 × 207 = 8534.61	The original question was 41.23 × 207. Use the estimated answer, 8000, to see where to place the decimal point. Alternatively, notice that 41.23 × 207 is 100 times less than 4123 × 207.	$4123 \times 207 = 853\,461$ $\qquad \div 100 \qquad\qquad \div 100$ $41.23 \times 207 = 8534.61$

Think about

Here is another method to work out where to put the decimal point in the answer to Worked example 5.

Count the number of digits after a decimal point in the question. There are two digits after the decimal point in 41.23 and no digits after the decimal point in 207, so the total is **two**. Place the decimal point **two** digits from the right of 853 461: the answer is 8534.61

$$41.\underline{23} \times 207 = 8534.\underset{\smile\smile}{\underline{61}}$$

Why does this method work? Does it always work?

Worked example 6

Estimate and then calculate $7.123 \div 3$. Round the answer to three decimal places (3 d.p.)

$7.123 \div 3 \approx 7 \div 3$ ≈ 2	Round so that can do the division in your head. 7.123 lies between 7 and 8, and it is closer to 7. $7 \div 3 = 2$ with remainder 1. You can ignore the remainder when making estimations.	7.123
$2 . 3\ 7\ 4\ 3...$ $3\ \overline{)\ 7\ .\ ^1 1\ ^2 2\ ^1 3\ ^1 0\ ^1}$	Write a decimal point in the answer above the decimal point in 7.123. To round the answer to 3 d.p., you need to know the fourth digit after the decimal point.	
$2.3743...$ $= 2.374$ (3.d.p.)	2.3743... lies between 2.374 and 2.375. The digit in the fourth decimal place is 3, so do not round up the 4.	(place value table: 1 • 0.1 0.01 0.001 0.0001 / 2 • 3 7 4 ③ / 2 • 3 7 4) 2.3743 2.374 ——— 2.375

Worked example 7

Estimate and then calculate $81.475 \div 25$. Write the exact answer.

$81.475 \div 25 \approx 80 \div 30$ ≈ 2		Round so that you can do the division in your head. 81.475 lies between 80 and 90, and it is closer to 80. 25 lies between 20 and 30. It is half-way between, so round it up to 30. $80 \div 30 = 2$ with remainder 20. You can ignore the remainder when you are estimating.	81.475 (between 80 and 90) 25 (between 20 and 30)
$\begin{array}{r} 3 . 2\ 5\ 9 \\ 25\overline{)\ 8\ ^8 1\ .\ 4\ 7\ 5} \\ 7\ 5\ \downarrow \\ 6\ 4 \\ 5\ 0 \\ 1\ 4\ 7 \\ 1\ 2\ 5 \\ 2\ 2\ 5 \\ 2\ 2\ 5 \\ 0 \end{array}$	25 50 75 100 125 150 175 200 225	Long division is shown on the left and short division is shown on the right. There is no remainder. $81.475 \div 25 = 3.259$	$\begin{array}{r} 3 . 2\ 5\ 9 \\ 25\overline{)\ 8\ ^8 1\ .\ ^6 4\ ^{14} 7\ 5} \end{array}$ 25 50 75 100 125 150 175 200 225

Exercise 4

1–6 and 8–9

1 Find the answers. You can use a written method.

a) 0.2×4 b) 9×0.06 c) 0.006×6 d) 3.007×8

e) $0.9 \div 3$ f) $0.56 \div 8$ g) $0.036 \div 2$ h) $0.036 \div 12$

2 For each calculation, estimate the answer by rounding each number.

Then calculate the exact answer using a written method.

a) 3.2×31 b) 7.32×41 c) -36×4.08

d) 9.16×83 e) $191.2 \div 5$ f) $-8.254 \div 2$

g) $100.008 \div 8$ h) $12.21 \div 11$

> **Tip**
>
> To estimate an answer, round each number so that you only need to use times tables and multiplication by powers of 10. For example,
> $4.78 \times 283 \approx 5 \times 300$
> $= 5 \times 3 \times 100$
> $= 1500.$

3 In the calculation ■ × ♦ = 0.072, ■ is a number with three digits after the decimal point and ♦ is an integer less than 13.

List the possible values of ■ and ♦.

4 For each calculation, estimate the answer by rounding each number first.

Then find the exact answer using a written method.

a) 63×78.5 b) 6.99×88 c) 11×1.111

d) 6.21×-506 e) $71.75 \div 25$ f) $54.94 \div 41$

g) $-147.52 \div 32$ h) $349.32 \div 71$

> **Tip**
>
> To estimate an answer, first round each number so that it has only one non-zero digit. For example, $472.81 \div 63 \approx 500 \div 60$. Then do the division, ignoring any remainder: $500 \div 60 \approx 8$ (since $8 \times 60 = 480$).

5 Owen says, 'It is not possible to calculate $5.62 \div 8$, because 8 is greater than 5.62.'

Is he correct? Explain your answer.

6 a) Use long division to calculate $78.44 \div 37$

b) Show the calculation $78.44 \div 37$ using short division.

c) Write one advantage of each method.

7 For each calculation, estimate the answer by rounding each number.

Then use a calculator to calculate the answer. Round each answer to three decimal places.

a) $33.82 \div 6$ b) $286.41 \div 44$ c) $-37.91 \div 27$ d) $55.891 \div 52$

8 Copy and complete the gaps in this long division.

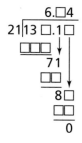

9 For each calculation, estimate the answer by rounding each number.

Then calculate the answer using a written method. Round each answer to the accuracy stated.

a) $22.058 \div 9$ (1 d.p.)

b) $13.569 \div 8$ (3 d.p.)

c) $9.0089 \div 8$ (4 d.p.)

d) $76.03 \div 33$ (2 d.p.)

e) $51.367 \div 21$ (2 d.p.)

f) $65.4321 \div 12$ (3 d.p.)

10 **Vocabulary question** Copy the sentences below and fill in the blanks using the list of words in the box.

quotient	divisor	remainder	dividend

In the calculation $72 \div 10$, the _____ is 72, the _____ is 10, the _____

is 7 and the _____ is 2.

 Thinking and working mathematically activity

'Lattice multiplication' is an alternative to long multiplication. Here is an example of lattice multiplication of 47 and 32:

The diagram shows the calculation $47 \times 32 = 1504$.

Work out the method for lattice multiplication, and try it with other integers. (Can you see why lattice multiplication works?)

Think about how to adapt the method for multiplying a decimal by an integer. How would you use lattice multiplication to calculate 4.7×32, 47×3.2, 0.47×32 or 47×0.32? Describe your method.

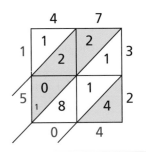

Consolidation exercise

1 State which of these numbers cannot be written as an integer power of ten.

100   11 000   10 000   100 000   10 000 000   101 000

2 Explain, using examples, whether each statement is always true, sometimes true, or never true.

a) When you multiply a number by 10, the answer will end in 0.

b) Dividing a number by 10 and then dividing the answer by 10 is the same as dividing the original number by 100.

c) Multiplying a number by 10^2 and then multiplying the answer by 10^3 is the same as multiplying the original number by 10 000.

3 Which of these calculations is the odd one out? Explain your answer.

a) 2.3×10^2

b) $23\,000 \div 10$

c) 0.23×10^3

d) 23×10

....
Tip
....
'Odd one out' means different to the others.
....

4 Using the ten numbers shown below, write them in the correct places to make all the statements true.

_____ rounded to the nearest integer is _____ .

_____ rounded to one decimal place is _____ .

_____ rounded to one decimal place is _____ .

_____ rounded to two decimal places is _____ .

_____ rounded to two decimal places is _____ .

23.488	23.049	23.48	24.0	23.999
23.0	24	23.482	23.951	23.49

5 Round each number to a suitable degree of accuracy, and explain your choice.

a) Jemma works out the mean number of people who attend her school's concert each year. The answer to her calculation is 245.816.

b) Faraz works out how much money each person gets if an amount of money is shared equally between five people. The answer to his calculation is $32.482.

6 Match the pairs of calculations that have the same answer.

−6.84 + 2.15
−2.15 + 6.84
−6.84 − 2.15

6.84 − 2.15
−(6.84 − 2.15)
−(6.84 + 2.15)

7 For each problem below, first estimate the answer. Then use any method (without a calculator) to find the exact answer.

a) Vesna has $56.22 in the bank.

She spends $82.45.

The amount of money Vesna has in the bank is now a negative number. Find this number.

b) On 1st November 2019, the lowest temperature recorded on Mars was −100.9 °C and the highest temperature was −24.6 °C.

Calculate the difference between the two temperatures.

8 Each block of the pyramid contains the sum of the two numbers below.

Use a written strategy to find the missing numbers.

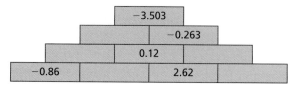

9 If 14.37 × 103 = 1480.11, which of these statements must also be true?

a) 1.437 × 103 = 148.011

b) 143.7 × 10.3 = 1480.11

c) 0.1437 × 1030 = 14.8011

d) 1480.11 ÷ 10.3 = 143.7

e) 148.011 ÷ 1437 = 0.0103

f) 103 ÷ 14.37 = 1480.11

10 Jenni multiplies a decimal number by a single-digit number. Her answer is 0.32. What could Jenni's calculation be? Give as many possibilities as you can.

End of chapter reflection

You should know that...	You should be able to...	Such as...
Multiplying by 10^n is the same as multiplying by ten n times.	Multiply integers and decimals by powers of 10.	Find: a) 52×10^3 b) 0.036×10^5
Dividing by 10^n is the same as dividing by ten n times.	Divide integers and decimals by powers of 10.	Find: a) $17 \div 10^4$ b) $12.79 \div 10^2$
To estimate the result of a calculation, round each number so that you can do the calculation mentally.	Estimate the results of adding and subtracting decimals. Estimate the result of multiplying or dividing a decimal by an integer.	Estimate: a) $4.68 + 7.93$ b) $62.35 - 86.194$ c) 2.354×27 d) $78.261 \div 22$
If you can see that an addition or a subtraction will have a negative answer, you can: • change the sign on each number • use a written method to do the calculation • write a negative sign on the result.	Add and subtract decimals using written methods, including when the answer is negative.	Calculate: a) $-3.28 + -4.51$ b) $22.4 - 38.07$ c) $-83.2 + 51.6$
To do long multiplication with an integer and a decimal, first ignore the decimal point and multiply. Then write the decimal point in the correct place in the answer.	Multiply a decimal by an integer.	Calculate: 17.82×107
To divide a decimal by an integer using written division, write a decimal point in the answer above the decimal point in the dividend.	Divide a decimal with two or more decimal places by a 1-digit or 2-digit integer.	Calculate: $7.384 \div 13$
When you do a division to a specified number of decimal places, work out one extra digit. Use this extra digit to see whether to round up the last digit in your answer.	Divide a decimal by an integer and write the answer to a specified number of decimal places.	Calculate: $36.85 \div 17$ and write the answer to three decimal places (3 d.p.)

8 Presenting and interpreting data 1

You will learn how to:

- Record, organise and represent categorical, discrete and continuous data.
  Choose and explain which representation to use in a given situation:
  - o Venn and Carroll diagrams
  - o tally charts, frequency tables and two-way tables
  - o dual and compound bar charts
  - o frequency diagrams for continuous data.
- Interpret data, identifying patterns, within and between data sets, to answer statistical questions. Discuss conclusions, considering the sources of variation, including sampling, and check predictions.

Starting point

Do you remember...

- how to record continuous and discrete data in a grouped frequency table?

  For example, copy and complete the following frequency table for a set of data.

Number of people	Tally	Frequency
0 – 4	III	
5 – 9	LHI	
10 – 14		11
15 – 19		2

- how to draw a frequency diagram for a simple grouped frequency table?

  For example, draw a frequency diagram to show the lengths of these insects.

Length (cm)	Number of insects
0 – 1	7
1 – 2	5
2 – 3	9
3 – 4	4

- how to interpret a Venn diagram?

  For example, the Venn diagram shows information about how 30 students got to school.

  How many students walked to school?

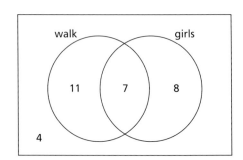

- how to draw and interpret a bar chart for a set of data?

For example, here is a bar chart showing how students in Year 8 travel to school.

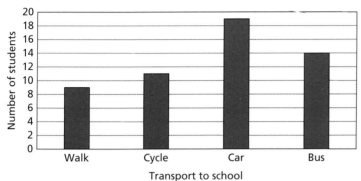

a) Write down the number of Year 8 students who travel by car.

b) Work out how many more Year 8 students travel by bus than cycle.

This will also be helpful when…

- you find averages from a frequency table
- you learn how to give the intervals in a grouped frequency table in a different way (as inequalities).

8.0 Getting started

Lata has drawn this bar chart to show the favourite fruit of students in her class.

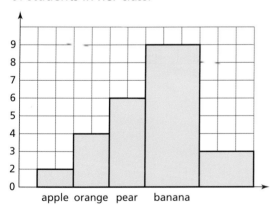

What is wrong with Lata's bar chart? Find as many errors as you can.

> **Think about**
>
> Can you draw a different statistical diagram with errors in it? Give your diagram to another student. Can they spot the errors you have included?

8.1 Two-way tables

Key terms

Two-way tables allow you to organise data for two different variables using columns and rows.

Worked example 1

A farmer keeps 18 sheep in a field. They are either white or black.

He has:

- 9 male sheep
- 10 black sheep
- 5 white female sheep

a) Construct and complete a two-way table to show this information.

b) What fraction of the farmer's black sheep are male?

> **Tip**
>
> When you have completed a two-way table, check carefully that all the rows and columns add up to the correct totals.

a)

	White	Black	Total
Male	3	6	9
Female	5	4	9
Total	8	10	18

Use the given information to work out the missing values.

For example, the missing total in the final column must be $18 - 9 = 9$.

b) The fraction of black sheep that are male is

$$\frac{6}{10} = \frac{3}{5}$$

Look at the 'black sheep' column.

There are 10 black sheep. 6 of these are male.

	White	Black	Total
Male	3	6	9
Female	5	4	9
Total	8	10	18

Exercise 1

1 Some dancers take part in a competition.

Each dancer chooses to perform either a piece of ballet or jazz.

They receive a mark between 1 and 10 for each dance.

The table shows the number of dancers getting each score.

	5 or less	6	7	8	9	10
Ballet	12	17	23	32	21	25
Jazz	6	26	41	23	7	7

a) How many of the dancers performing jazz scored a mark of 6?

b) How many more dancers performed a piece of ballet than jazz?

c) What fraction of all the dancers scored a mark of at least 9?

2 28 musicians are taking a practical exam.

The two-way table gives information about the instruments that the musicians have decided to use for their practical exam.

	Male	Female	Total
Voice	1	3	
Piano	2		
Guitar		1	6
Clarinet		2	2
Percussion	4		5
Saxophone	1	3	
Total			

a) Copy and complete the two-way table.

b) How many students decided to play the piano?

3 The two-way table gives information about the gender and ages of the members of a running club.

Age (years)	11–20	21–30	31–40		51–60	Over 60
Male	5	3	12	14	3	2
Female	7	2	15	11	5	3

a) What is the missing age range in the table?

b) How many members of the running club are female?

c) How many members of the running club are male and have an age between 21 and 30 years old?

d) How many members of the running club have an age of at least 51 years?

e) The leader of the running club says that more than half of the members of the running club are aged 41 years or older.

Is the leader correct? Show how you worked out your answer.

4 80 students from two schools were asked how they usually travel to school.

	Walk	Car	Bus	Total
School A				
School B				
Total				80

Equal numbers of students were asked from each of the two schools.

20% of the 80 students travel by bus. 10 of these students who travel by bus were from School A.

14 students from School A walked to school.

Half as many students from School B travel by car as travel by car from School A.

a) Use the information to copy and complete the two-way table.

b) Show that 30% of the 80 students usually travel by car.

5 The table and the Venn diagram show information about the number of hotel customers who ate breakfast and dinner on two days.

Monday

	Ate breakfast	Did not eat breakfast
Ate dinner	30	6
Did not eat dinner	33	17

Tuesday

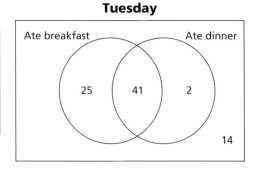

a) Find the total number of hotel customers on Monday.

b) How many customers ate both breakfast and dinner on Tuesday?

c) Calculate how many more customers ate dinner on Tuesday than on Monday.

6 George collects model cars and vans. His models are either old or new.

a) Copy and complete the two-way table.

b) Copy and complete the Venn diagram to show his data.

	Old	New	Total
Cars			45
Vans	6		
Total	23		60

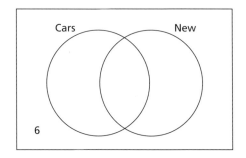

7 On a bus there are 38 students from two different schools:

Heartshead High and Staley Farm College.

7 of the 18 students who go to Heartshead High are male.

There are 8 female students who go to Staley Farm College.

In total, how many students on the bus are male?

Tip

Put the information from the question into a two-way table.

8 A company makes and sells bracelets.

They sell the bracelets in a shop and online.

The table shows the sale of bracelets each month in 2019.

Month	Jan	Feb	Mar	Apr	May	Jun	Jul	Aug	Sep	Oct	Nov	Dec
Sales in the shop	340	115	120	109	132	127	137	97	104	117	290	484
Sales online	134	145	146	136	138	130	143	102	111	118	135	155

a) The manager makes this conclusion from the table:

In most months in 2019, the company sold more bracelets online than it did in the shop.
Comment on the manager's conclusion.

b) Write a different conclusion that can be made from the table.

▼▼ Thinking and working mathematically activity

Karim owns a small supermarket. He records the number of items bought by 30 male and by 30 female customers.

Male cutomers
2 19 8 3   9 3 8 2 14 20
3 4 6 12 5 4 6 9 3 2
4 8 10 2 18 11 3 9 2 4

Female customers
7 9 28 18 10 8 19 2 11 6
17 12 8 7 26 4 8 3 17 25
14 4 3 6 15 10 16 9 2 10

- Give a reason why a two-way table is an appropriate way to summarise this information.
- Create and complete your own two-way table for the data.
- Write some conclusions about the number of items bought by the male and female customers.
- Compare your table with the tables produced by other students. Are some tables more effective than others?

8.2 Dual and compound bar charts

Key terms

Dual bar chart: A bar chart for showing data from two groups side by side.

Compound bar chart: A bar chart showing data for two or more groups with the bars stacked on top of each other.

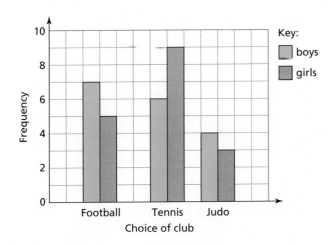

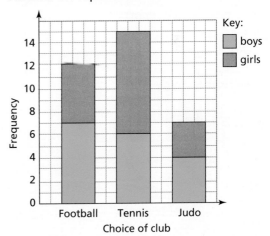

Discuss

What do the graphs on page 87 show about the choices of club made by these children?

Can you think of any advantages or disadvantages of each type of graph?

Did you know?

Compound bar charts are sometimes known as stacked bar charts.

Worked example 2

The incomplete dual bar chart shows some information about the favourite types of television show for two classes in Year 8: class 8r and class 8s.

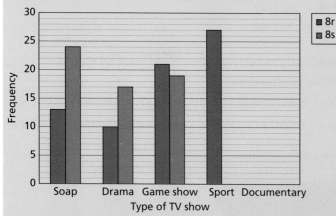

The number of class 8s who said that 'Sport' was their favourite type of television show was one third of the number of class 8r who said that 'Sport' was their favourite type of television show.

9 of class 8r and 11 of class 8s said that 'Documentary' was their favourite type of television show.

a) Complete the bar chart.

b) Write down the most popular type of television show for class 8s.

a) 27 of class 8r said 'Sport'.

$27 \div 3 = 9$

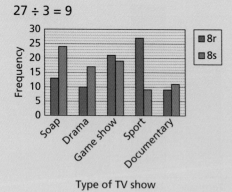

Start by working out how many of class 8s said that 'Sport' was their favourite type of television show.

You need to find $\frac{1}{3}$ of the number of class 8r who said 'Sport'.

You need to add the bars for students in class 8s who said 'Sport' and for 'Documentary' to the bar chart.

Tip

Remember to use the key to check the colour for the bars for class 8r and for class 8s.

b) Looking at the orange bars, you can see that 'Soap' has the tallest bar.

'Soap' is the most popular type of television show for class 8s.

The key shows you that the bars for class 8s are orange. To find the most popular type of television show for class 8s you need to find the tallest orange bar.

1 Draw a dual bar chart to show the data in each table.

a)

Instrument played

	Piano	Violin	Drum
Boys	11	7	2
Girls	9	4	7

b)

Drinks sold in a café one day

	Coffee	Tea	Juice	Other
Morning	10	5	6	3
Afternoon	4	11	5	6

2 a) Draw a compound bar chart to illustrate the data in question 1 **a)**.
The horizontal axis should be the type of instrument played.

b) Draw a compound bar chart to illustrate the data in question 1 **b)**.
The horizontal axis should be the type of drink sold.

3 The dual bar chart shows the favourite colour of all the boys and girls in Year 7.

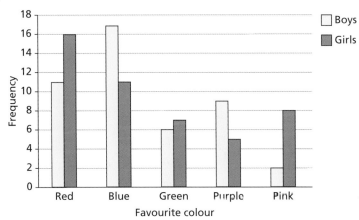

a) Write down the number of girls whose favourite colour is green.

b) Write down the difference between the number of boys and the number of girls giving purple as their favourite colour.

c) Find how many more girls than boys there are in Year 7.

d) The school is planning to repaint the Year 7 social area using one of these colours.
Use the dual bar chart to decide which colour they should use. Explain your answer.

4 The compound bar chart shows the number of students studying some languages.

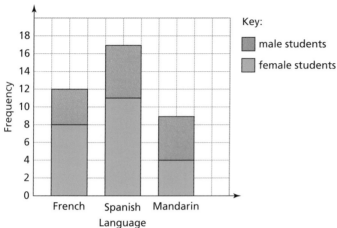

a) Write down the language that is studied by the greatest number of students.

b) Write down how many female students study Spanish.

c) Find how many more male students than female students study Mandarin.

d) Juan says, 'There are more female students than male students.'

   Is Juan correct? Give a reason for your answer.

5 The compound bar chart shows the types of books borrowed from a library one morning.

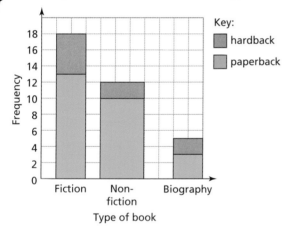

a) Write down the problems with this compound bar chart.

b) Draw the diagram correctly.

6 The dual bar chart shows the number of adults and children watching a film on three days.

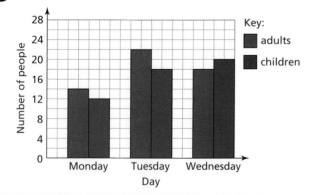

a) How many people watched the film on Monday?

b) Compare the number of adults who watched the film on Tuesday with the number of children who watched on Tuesday.

c) Give a reason why a dual bar chart is a suitable type of graph to show the information.

Thinking and working mathematically activity

Real data question The compound bar chart shows the population of the world in 2000, 2008 and 2016. It also shows the number of people with and without clean drinking water.

Key:
- people without clean drinking water
- people with clean drinking water

Think about

Use the information in the graph to make a prediction for the population of the world in 2024. How many people do you think will have clean drinking water in 2024? Give reasons for your predictions.

Source: The World Bank: People using safely managed drinking water services (% of population): WHO/UNICEF Joint Monitoring Programme (JMP) for Water Supply, Sanitation and Hygiene (washdata.org).

Use the information in the graph to make some conclusions.

8.3 Frequency diagrams for continuous data

Worked example 3

A park has 264 trees. Pierre measures the height (in metres) of a random sample of 60 trees.

a) How many of the trees in Pierre's sample had a height under 20 metres?

b) What fraction of the trees in Pierre's sample were over 40 metres tall?

c) Estimate the number of trees in the whole park that are over 40 metres tall.

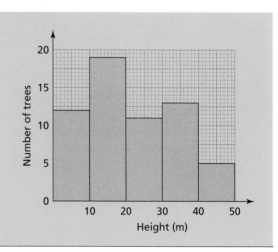

a) The number of trees with height under 20 metres is

12 + 19 = 31

The first two bars in the diagram correspond to trees with a height of under 20 metres.

The frequencies for these two bars are 12 and 19.

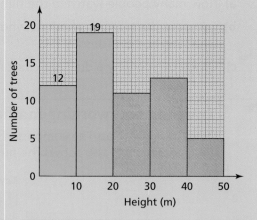

b) 5 out of the 60 trees in the sample were over 40 metres tall.

The fraction is:

$$\frac{5}{60} = \frac{1}{12}$$

The last bar corresponds to trees over 40 metres tall.

The frequency for this bar is 5.

The total number of trees in the sample is 60.

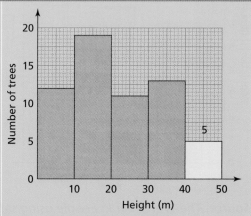

c) $\frac{1}{12}$ of 264 = 264 ÷ 12 = 22

So about 22 trees in the park should have a height of over 40 metres.

The fraction of all trees in the park that are over 40 metres tall should be approximately the same as in the sample.

So, about $\frac{1}{12}$ of the 264 trees in the park should be over 40 metres tall.

1 in 12 trees are over 40 metres tall.

Exercise 3

1 Use graph paper to draw frequency diagrams for each set of data.

a) Speeds of some planes

Speed (km/h)	Frequency
700 – 750	11
750 – 800	17
800 – 850	36
850 – 900	28
900 – 950	15

b) Lengths of some giant pandas

Length (m)	Frequency
1.1 – 1.2	13
1.2 – 1.3	29
1.3 – 1.4	43
1.4 – 1.5	35
1.5 – 1.6	15

2 Sergei records the mass (in kilograms) of suitcases taken on board a plane.

19.5 20.4 21.8 30.7 24.8 23.1 23.4 27.5 20.9 29.7 18.4 31.8 23.8 29.7

28.5 21.3 24.5 23.7 28.1 29.0 27.4 28.2 21.1 20.8 23.2 28.7 25.2 21.5

30.2 30.8 24.7 27.3

a) Create a frequency table to summarise Sergei's data. Your table should have equal width intervals.

b) Use your table to draw a frequency diagram.

> **Tip**
>
> An interval written 18 – 20 kg would usually represent masses from 18 kg up to (but not including) 20 kg.

3 The table shows the time it took some children to finish a task.

	Time (minutes)				
	10 – 15	15 – 20	20 – 25	25 – 30	30 – 35
Boys	19	27	41	33	16
Girls	11	15	32	41	21

a) Draw a frequency diagram to show the time it took the boys.

b) Draw a separate frequency diagram to show the time it took the girls.

c) Give a reason why frequency diagrams are appropriate diagrams for showing the data.

4 Pedro draws a frequency diagram to show the height of some children.

 a) Write down what is wrong with how Pedro has drawn his diagram.

 b) Draw Pedro's frequency diagram correctly.

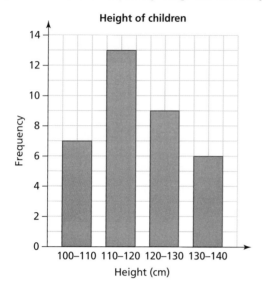

5 The frequency diagram shows the time when the first goal was scored in matches in a school football tournament

(Matches where no goals were scored or where the goal was scored in extra time are not included.)

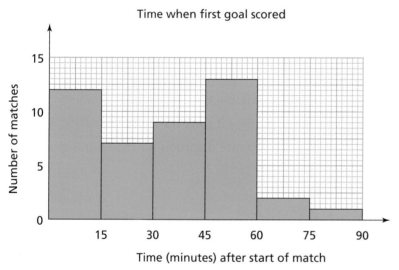

 a) Write down the number of matches when the first goal was scored within the first 30 minutes.

 b) A football match has two halves of 45 minutes each. Find the number of matches where the first goal was not scored until the second half. Give a reason for your answer.

6 A town has a population of 20 000. Jim records the age of a random sample of people in the town. The frequency diagram shows his results.

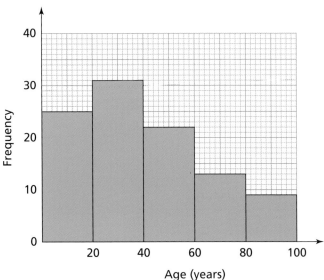

a) How many people in Jim's sample were more than 60 years old?

b) What fraction of the people in Jim's sample were under 20 years old?

c) Estimate the number of people in the whole town who are under 20 years old.

7 Archie and Sophia both take a random sample of 120 cabbages growing in a field.

The frequency diagrams show their results.

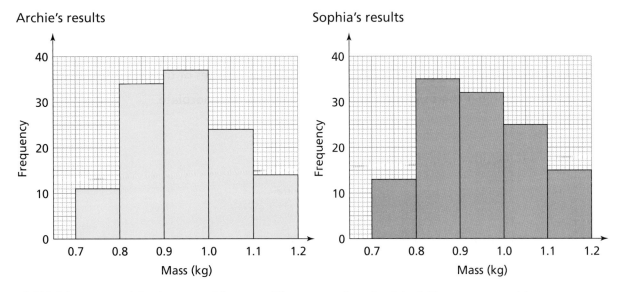

a) Which person picked more cabbages with a mass of under 1 kg? Show your working.

b) Show that one eighth of Sophia's cabbages have a mass of more than 1.1 kg.

c) The farmer is concerned that Archie's results are not identical to Sophia's results.

He thinks that either Archie or Sophia must have recorded their masses incorrectly.

Could the farmer be correct? Give a reason for your answer.

Thinking and working mathematically activity

Make a conjecture about the length of popular songs.

The data below shows the length (seconds) of 45 popular songs.

336	159	208	219	156	159	184	189	210
227	221	166	208	163	245	193	205	227
229	187	160	201	213	193	131	189	158
188	207	115	197	159	196	200	237	203
221	211	332	167	213	212	168	174	214

- Create and complete a frequency table to summarise the data.
- Draw a frequency diagram to show the data.
- Make some conclusions. Do the data support your conjecture?

Consolidation exercise

1 Brian draws the following two-way table to show the favourite flavour of crisps of a group of his friends. He made a mistake in the table.

	Prawn cocktail	Cheese and onion	Salt and vinegar	Ready salted	Total
Male	2	5	6	8	21
Female	4	4	8	7	24
Total	6	9	14	15	45

He knows that exactly $\frac{1}{3}$ of his friends prefer salt and vinegar. Find the two values which are incorrect in the table.

2 A box contains milk chocolates and plain chocolates. The centres of the chocolates are soft or hard. The compound bar chart shows the number of each type of chocolate.

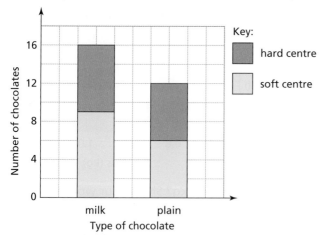

Decide if each statement is true or false.

a) Sixteen of the chocolates are milk chocolates with hard centres.

b) There are more milk chocolates than plain chocolates.

c) More of the chocolates have hard centres than soft centres.

3 The students at a small school are given a piece of fruit with their lunch.

Students can choose either an apple or a banana.

There are 80 students at the school.

50 of the students are female.

Three-quarters of the students chose a banana.

10% of the students who chose an apple are male.

a) Copy and complete the table to show the information.

	Apple	Banana	Total
Male			
Female			
Total			

b) Draw a dual bar chart to show the information.

4 The table shows the length of some caterpillars.

Length (mm)	Frequency
10 – 20	17
20 – 30	25
30 – 40	38
40 – 50	34
50 – 60	19

Draw a frequency diagram to show the data.

5 The frequency diagram shows the time it took some runners to complete a race.

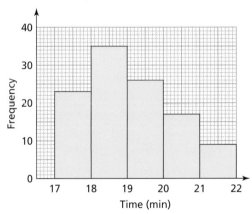

a) Find the number of runners that completed the race in under 20 minutes.

b) Jack is one of these runners. He claims he finished the race in 16 minutes 30 seconds.

Explain why he must be incorrect.

c) Tilly says that the slowest runner must have taken 22 minutes to run the race.
Explain why she might be incorrect.

End of chapter reflection

You should know that...	You should be able to...	Such as...
A two-way table shows data about two different variables. One variable is shown in the rows and the other in the columns.	Organise data into a two-way table. Interpret data presented as a two-way table.	Copy and complete this two-way table. | Drink	Adults	Children	Total
---	---	---	---
Juice		4	
Fizzy			31
Total	24		45	
A dual bar chart is a type of bar chart which has two bars drawn next to each other. A compound bar chart is also a type of bar chart. Each bar is divided into different groups.	Draw and interpret a dual bar chart and a compound bar chart.	Draw a dual bar chart and a compound bar chart to show the data in the table above. Plot type of drink on the horizontal axis.
Grouped data can be shown in a frequency diagram. For grouped continuous data, the bars in a frequency diagram should be drawn without gaps.	Draw and interpret a frequency diagram for grouped continuous data.	 Length of animals (Frequency vs Length (cm)) How many animals were longer than 20 cm long?

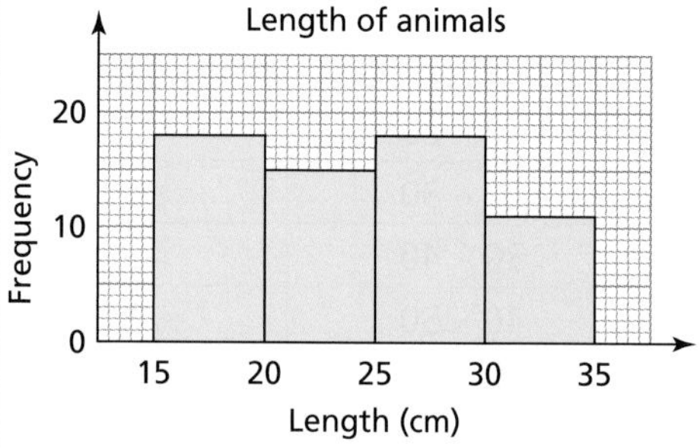

9 Fractions

You will learn how to:
- Estimate and add mixed numbers, and write the answer as a mixed number in its simplest form.
- Estimate, multiply and divide proper fractions.

Starting point

Do you remember...

- how to simplify a fraction?

  For example, simplify $\frac{28}{70}$
- how to add fractions with different denominators?

  For example, find $\frac{3}{8} + \frac{7}{12}$
- how to convert between improper fractions and mixed numbers?

  For example, write $\frac{17}{3}$ as a mixed number, and write $2\frac{3}{4}$ as an improper fraction.
- how to add improper fractions with the same or different denominators?

  For example, find $\frac{7}{2} + \frac{5}{4}$
- how to multiply a proper fraction by an integer?

  For example, find $9 \times \frac{2}{3}$
- how to divide a proper fraction by an integer?

  For example, find $\frac{3}{5} \div 3$

This will also be helpful when...

- you learn to subtract mixed numbers
- you learn to multiply an integer by a mixed number
- you learn to divide an integer by a proper fraction.

9.0 Getting started

At a big party, there are lots of cakes. The cakes are identical except that some are blue and some are yellow. The numbers in the diagram show the fractions of cakes (and also some whole cakes) left over after the party. Jörgen takes home all of the leftover blue cakes. Kiana takes home all of the leftover yellow cakes.

Which sum is greater? Find these answers without a calculator.

$$1 + \frac{8}{9} + \frac{2}{10} + \frac{4}{5} + \frac{1}{6} + \frac{5}{6} + \frac{1}{7} + \frac{6}{7} + \frac{1}{8} + \frac{1}{9} + \frac{7}{8} = ?$$

$$4 + \frac{1}{2} + \frac{1}{4} + \frac{1}{7} + \frac{1}{14} + \frac{1}{28} = ?$$

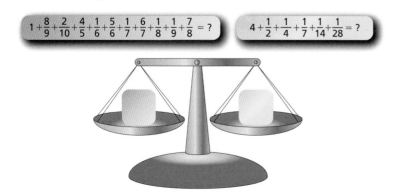

Tip

Look for ways to make the calculation as easy as possible! You could rearrange the fractions before doing each calculation.

Key terms

A **proper fraction** is a fraction where the numerator is less than the denominator. For example, $\frac{1}{4}$ and $\frac{7}{9}$ are both proper fractions.

An **improper fraction** is a fraction where the numerator is greater than the denominator. For example, $\frac{5}{4}$ and $\frac{22}{5}$ are both improper fractions.

A **mixed number** is a number made of an integer and a proper fraction. For example, $1\frac{1}{4}$ and $4\frac{2}{5}$ are both mixed numbers.

> **Did you know?**
>
> The word fraction comes from the Latin word 'fractio' meaning 'breaking'.

Worked example 1

Estimate $2\frac{1}{3} + 1\frac{4}{5}$, and then calculate the exact answer.

$2\frac{1}{3} + 1\frac{4}{5} \approx 2 + 2$ $= 4$	To estimate the answer, round each mixed number to the nearest integer. $2\frac{1}{3}$ is closer to 2 than to 3. $1\frac{4}{5}$ is closer to 2 than to 1.	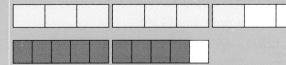
$\frac{1}{3} + \frac{4}{5} = \frac{5}{15} + \frac{12}{15}$ $= \frac{17}{15}$ $= 1\frac{2}{15}$ $2\frac{1}{3} + 1\frac{4}{5} = 1\frac{2}{15} + 2 + 1$ $= 4\frac{2}{15}$	Add the fractional parts. Start by writing both fractions with a common denominator. The lowest common denominator is 15. Add the integers.	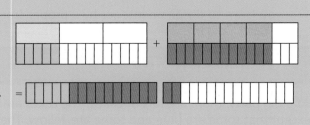

Exercise 1   1–5, 7–8

1 Asha is adding two fractions. Here is her working:

$$\frac{3}{8} + \frac{1}{4} = \frac{4}{12} = \frac{1}{3}$$

Give a reason why Asha's working is not correct.

2 Estimate the answers to these additions, then calculate the exact answers.
Give each answer as a mixed number in its simplest form.

a) $2\frac{1}{5} + \frac{3}{5}$

b) $3\frac{1}{2} + \frac{5}{12}$

c) $1\frac{1}{9} + 1\frac{2}{3}$

d) $\frac{1}{3} + 1\frac{7}{10}$

e) $1\frac{3}{5} + 2\frac{2}{3}$

f) $3\frac{2}{3} + 1\frac{5}{7}$

g) $2\frac{2}{15} + 1\frac{3}{5}$

h) $4\frac{7}{8} + 2\frac{2}{3}$

3 Match the pairs of calculations that have the same answer.

A $1\frac{5}{12} + 3\frac{1}{3}$

B $3\frac{1}{2} + 1\frac{7}{12}$

C $2\frac{2}{3} + 2\frac{1}{4}$

D $3\frac{1}{12} + 1\frac{5}{6}$

E $2\frac{7}{8} + 1\frac{7}{8}$

F $3\frac{5}{6} + 1\frac{1}{4}$

4 A running track is $3\frac{1}{2}$ kilometres long. I walk for $2\frac{3}{10}$ kilometres and run for $1\frac{1}{5}$ kilometres.
Will I complete the full distance of the track?

5 Find the missing values:

a) $\boxed{}\dfrac{\boxed{}}{\boxed{}} - 2\frac{5}{6} = 4\frac{1}{4}$

b) $\boxed{}\dfrac{\boxed{}}{\boxed{}} - 1\frac{3}{10} = 3\frac{5}{8}$

> **Tip**
> If you knew $\boxed{} - 5 = 2$, how would you find $\boxed{}$?

6 **Technology question** Find out how to enter a mixed number into your calculator.

Estimate the answer to each calculation below.

Then use your calculator to do the calculation.

a) $2\frac{2}{9} + \frac{5}{11}$

b) $\frac{1}{8} + 2\frac{3}{4}$

c) $3\frac{4}{5} + 1\frac{6}{7}$

d) $13\frac{3}{8} + 5\frac{11}{12}$

7 Use the digits 1, 2, 3, 4, 5 and 6 once each to make this calculation correct:

$$\boxed{}\dfrac{\boxed{}}{\boxed{}} + \boxed{}\dfrac{\boxed{}}{\boxed{}} = 4\frac{7}{12}$$

8 Mary-Jo and Baljit calculate $2\frac{3}{4} + 1\frac{7}{12}$. They both get the correct answer.

a) Criticise their methods, describing any disadvantages.

Mary Jo $2\frac{3}{4} + 1\frac{7}{12} = 3 + \frac{36}{48} + \frac{28}{48}$

$= 3\frac{64}{48}$

$= 4\frac{16}{48}$

$= 4\frac{1}{3}$

Baljit $2\frac{3}{4} + 1\frac{7}{12} = \frac{11}{4} + \frac{19}{12}$

$= \frac{33}{12} + \frac{19}{12}$

$= \frac{52}{12}$

$= \frac{13}{3}$

$= 4\frac{1}{3}$

> **Think about**
> When you add two fractions, if you choose a common denominator that is not the lowest common denominator, do you still get the correct answer? What difference does it make?

b) Show a more efficient method for calculating $2\frac{3}{4} + 1\frac{7}{12}$

Thinking and working mathematically activity

Investigate the results of these calculations:

$$\frac{1}{1} + \frac{1}{2} =$$

$$\frac{1}{1} + \frac{1}{2} + \frac{1}{4} =$$

$$\frac{1}{1} + \frac{1}{2} + \frac{1}{4} + \frac{1}{8} =$$

$$\frac{1}{1} + \frac{1}{2} + \frac{1}{4} + \frac{1}{8} + \frac{1}{16} =$$

Describe any patterns you see in the results.

Predict, without doing any calculation, what the next three results will be.

Try to explain the patterns.

Will the sum ever be greater than 2? Explain your answer.

9.2 Multiplying and dividing proper fractions

Key terms

To **invert** a fraction is to swap the numerator and denominator. For example, if you invert $\frac{5}{8}$ you get $\frac{8}{5}$.

The **reciprocal** of a number equals 1 divided by the number. The product of a number and its reciprocal is 1.

For example, the reciprocal of 2 is $1 \div 2 = \frac{1}{2}$. (Notice that $2 \times \frac{1}{2} = 1$). You can find the reciprocal of a fraction by inverting it. For example, the reciprocal of $\frac{2}{3}$ is $\frac{3}{2}$. (Notice that $\frac{2}{3} \times \frac{3}{2} = 1$).

> **Think about**
>
> Is each statement below always true, sometimes true or never true?
>
> If a number is less than 1, then its reciprocal is greater than 1.
>
> If a number is greater than 1, then its reciprocal is less than 1.

Worked example 2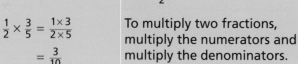

For each calculation, first state whether the answer is bigger or smaller than the first number. Then calculate the exact answer.

a) $\frac{1}{2} \times \frac{3}{5}$ b) $\frac{3}{4} \times \frac{5}{12}$ c) $\frac{1}{3} \div \frac{1}{6}$ d) $\frac{3}{8} \div \frac{3}{4}$

a) $\frac{1}{2} \times \frac{3}{5} < \frac{1}{2}$

If you multiply a number by a fraction that is less than 1, the result is smaller than that number. So $\frac{3}{5}$ of $\frac{1}{2}$ is smaller than $\frac{1}{2}$.

$$\frac{1}{2} \times \frac{3}{5} = \frac{1 \times 3}{2 \times 5}$$

$$= \frac{3}{10}$$

To multiply two fractions, multiply the numerators and multiply the denominators.

b) $\frac{3}{4} \times \frac{5}{12} < \frac{3}{4}$

$\frac{5}{12}$ is smaller than 1 so $\frac{5}{12}$ of $\frac{3}{4}$ is less than $\frac{3}{4}$.

$\frac{^1\cancel{3}}{4} \times \frac{5}{\cancel{12}_4} = \frac{1}{4} \times \frac{5}{4}$
$= \frac{5}{16}$

Before multiplying fractions, check whether you can cancel first.

You can cancel by dividing any numerator and any denominator by the same number.

Here you can divide by 3.

Alternatively you can multiply first and then cancel.

$3 \div 3 = 1$

$\frac{\cancel{3}}{4} \times \frac{5}{\cancel{12}} = \frac{1}{4} \times \frac{5}{4}$

$12 \div 3 = 4$

$\frac{3}{4} \times \frac{5}{12} = \frac{15}{48}$
$= \frac{5}{16}$ $\Big)\div 3$

c) $\frac{1}{3} \div \frac{1}{6} = \frac{1}{3} \times 6$

$\frac{1}{3} \times 6 > \frac{1}{3}$

Dividing by $\frac{1}{6}$ is the same as multiplying by 6. This makes the answer bigger than $\frac{1}{3}$.

Dividing a number by a fraction less than 1 always gives an answer bigger than the number (because it is equivalent to multiplying by a fraction greater than 1).

$\frac{1}{3} \div \frac{1}{6} = \frac{1}{3} \times 6$
$= 2$

Dividing by a number gives the same result as multiplying by its reciprocal.

The reciprocal of $\frac{1}{6}$ is 6.

You can subtract $\frac{1}{6}$ from $\frac{1}{3}$ exactly 2 times:

d) $\frac{3}{8} \div \frac{3}{4}$

$\frac{3}{8} \div \frac{3}{4} > \frac{3}{8}$

$\frac{3}{4}$ is less than 1.

Dividing by a fraction less than 1 will give an answer that is bigger than $\frac{3}{8}$.

$\frac{3}{8} \div \frac{3}{4} = \frac{3}{8} \times \frac{4}{3}$
$= \frac{1}{2} \times \frac{1}{1} = \frac{1}{2}$

Dividing by $\frac{3}{4}$ gives the same result as multiplying by $\frac{4}{3}$.

Before multiplying, check whether you can cancel first. You can divide by 3 and also by 4.

$\frac{3}{4} = \frac{6}{8}$, so the question is

$\frac{3}{8} \div \frac{6}{8}$ = 3 eighths ÷ 6 eighths = $3 \div 6 = \frac{1}{2}$.

Exercise 2

1–5, 8–9

1 For each calculation, first state whether the answer will be bigger or smaller than the first number. Then find the exact answer.

Write fractional answers as proper or improper fractions in their simplest form.

a) $3 \times \frac{2}{7}$

b) $\frac{2}{3} \times 4$

c) $\frac{3}{8} \times 8$

d) $15 \times \frac{4}{5}$

e) $\frac{3}{4} \div 3$

f) $\frac{8}{9} \div 2$

g) $\frac{3}{4} \div \frac{1}{4}$

h) $\frac{2}{5} \div 6$

2 Write your answers to these in their simplest form.

a) $\frac{1}{3} \times \frac{1}{4}$

b) $\frac{2}{5} \times \frac{1}{8}$

c) $\frac{3}{4} \times \frac{5}{9}$

d) $\frac{8}{9} \times \frac{2}{3}$

e) $\frac{8}{9} \times \frac{3}{4}$

f) $\frac{3}{10} \times \frac{5}{7}$

g) $\frac{2}{5} \times \frac{2}{5}$

h) $\frac{7}{8} \times \frac{4}{21}$

3 Write the reciprocal of each number.

a) $\frac{3}{4}$

b) $\frac{9}{2}$

c) $\frac{1}{2}$

d) 3

4 Write your answers to these in their simplest form.

a) $\frac{4}{7} \div \frac{1}{7}$

b) $\frac{8}{9} \div \frac{4}{9}$

c) $\frac{3}{5} \div \frac{1}{10}$

d) $\frac{3}{10} \div \frac{1}{5}$

e) $\frac{8}{9} \div \frac{3}{4}$

f) $\frac{7}{12} \div \frac{5}{7}$

g) $\frac{2}{3} \div \frac{1}{8}$

h) $\frac{10}{11} \div \frac{9}{22}$

> **Discuss**
>
> If you multiply two proper fractions, is the answer always, sometimes or never smaller than each of the two fractions? Explain your answer.

5 Carlos works out $\frac{3}{4} \div \frac{5}{9}$. His working is shown below.

$$\frac{\overset{1}{\cancel{3}}}{4} \div \frac{5}{\underset{3}{\cancel{9}}} = \frac{1}{4} \times \frac{3}{5}$$

$$= \frac{3}{20}$$

a) Describe the incorrect step Carlos has taken.

b) Explain why this step is not valid.

c) Show how to do the calculation correctly.

6 For each calculation below, first write whether the answer will be greater than or less than the first number.

Then use a calculator and write your answers as proper or mixed fractions.

a) $\frac{13}{14} \times \frac{13}{14}$

b) $\frac{11}{20} \times \frac{3}{8}$

c) $\frac{1}{15} \div \frac{21}{22}$

d) $\frac{24}{25} \div \frac{2}{15}$

7 Technology question Find out where the reciprocal key is on your calculator.

a) Use this key to find the reciprocal of the reciprocal of $\frac{1}{2}$.

b) Explain why the answer has this value.

8 Ben writes $\frac{1}{4} \times \frac{2}{3} = \frac{3}{8}$. Explain how you know that this answer cannot be correct.

Calculate the correct answer.

Thinking and working mathematically activity

Investigate the questions below. For each question, you could start by looking for examples that agree or disagree with the statement. Then look for patterns, and try to explain them.

When you divide a positive proper fraction by a proper fraction:

- is the answer always, sometimes or never larger than 1?
- is the answer always, sometimes or never larger than the first fraction?
- is the answer always, sometimes or never larger than the second fraction?

Consolidation exercise

1 Estimate the answer to each calculation. Then calculate the exact answer, giving the answer as an improper fraction.

a) $2\frac{1}{3} + 1\frac{1}{5}$　　b) $3\frac{3}{4} + \frac{5}{8}$　　c) $1\frac{5}{8} + 1\frac{1}{6}$　　d) $4\frac{4}{5} + 3\frac{1}{10}$

2 Amélie watched two films. The first film was $1\frac{3}{4}$ hours long. The second film was $2\frac{2}{5}$ hours long.

Find the total length of the two films, giving your answer as a mixed number.

3 Estimate the answer to each calculation. Then calculate the exact answers, giving each answer as an improper fraction.

a) $1\frac{1}{8} + 3\frac{5}{16} + 2\frac{1}{4}$　　　　b) $1\frac{1}{5} + 1\frac{3}{20} + 2\frac{1}{2}$

4 Gunnar and Lowanna both calculate $\frac{5}{18} \times \frac{9}{15}$. They both get the correct answer.
Their methods are shown below.

Gunnar

$\frac{5}{18} \times \frac{9}{15} = \frac{45}{270}$

$= \frac{9}{54}$

$= \frac{1}{6}$

Lowanna

$\frac{5}{18} \times \frac{9}{15} = \frac{1}{2} \times \frac{1}{3}$

$= \frac{1}{6}$

a) Describe Gunnar's first step.

b) Describe Lowanna's first step.

c) Do you think one method is better than the other? Explain your answer.

5 Josh spends $\frac{3}{4}$ hour doing maths homework. He spends $\frac{2}{5}$ of this time practising fraction multiplication.

Calculate what fraction of an hour he spends practising fraction multiplication.

End of chapter reflection

You should know that...	You should be able to...	Such as...
To add mixed numbers, you can add the integer parts and the fraction parts separately, and then combine the results.	Add two mixed numbers.	Find: $3\frac{1}{10} + 2\frac{5}{8}$
To multiply two fractions, multiply the numerators and multiply the denominators. You can cancel before or after multiplying.	Multiply two proper fractions.	Find: $\frac{2}{3} \times \frac{6}{7}$
When you multiply a number by a proper fraction, the result is less than the number. When you divide a number by a proper fraction, the result is greater than the number.	Estimate whether the result of multiplying a number by a fraction less than 1 will be greater or less than the number. Estimate whether the result of dividing a number by a fraction less than 1 will be greater or less than the number.	State whether $\frac{2}{5} \times \frac{7}{9}$ is greater or less than $\frac{2}{5}$. State whether $\frac{2}{5} \div \frac{7}{9}$ is greater or less than $\frac{2}{5}$.
To divide a fraction by a fraction, invert the second fraction and multiply.	Divide a proper fraction by a proper fraction.	Find: $\frac{4}{9} \div \frac{5}{6}$

10 Manipulating expressions

You will learn how to:

- Understand how to manipulate algebraic expressions, including:
  - collecting like terms
  - applying the distributive law with a constant.

Starting point

Do you remember...

- how to write an algebraic expression to represent something in real-life?

  For example, you think of a number. You double the number and then add 5. You can write the new number as $2m + 5$, where m represents your original number.

- how to multiply by 2-digit whole numbers?

  For example, $3 \times (42) = 3 \times 40 + 3 \times 2 = 120 + 6 = 126$

This will also be helpful when...

- you learn about constructing and using formulas.

10.0 Getting started

Eva is organising a charity cake stall. She has five boxes of cupcakes with c cakes in each box, plus another 12 cupcakes.

How could you write this as an algebraic expression?

She is now given another eight boxes of cupcakes.

How many cupcakes does she have now? Write an algebraic expression.

She sells 12 whole boxes of cupcakes and another seven separate cupcakes.

How many cupcakes does she have left?

If there were eight cupcakes in each box, how many cupcakes did she sell?

How many cupcakes did she have left?

Can you make up a problem like this?

10.1 Simplifying expressions

Key terms

When more than one term with the same letter is added or subtracted it is called **collecting like terms** or **simplifying by collecting like terms**.

For example, a rectangle has side lengths c and d. The perimeter of the rectangle is

$c + d + c + d$

As you have two lots of c and two lots of d, you can write this as

$2c + 2d$

In algebra, $2 \times d$ is written as $2d$. This is to make sure that you don't confuse a multiplication sign with the letter x. Also, the number is always written before the letter.

Worked example 1

a) Form two expressions for the total mass of these five boxes:

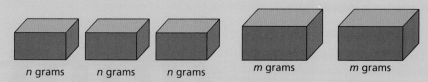

n grams n grams n grams m grams m grams

b) Form an expression for the perimeter of the triangle shown.

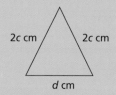

$2c$ cm $2c$ cm d cm

c) Simplify $m + n + 3m + 2n - 2m$

d) Simplify $4 \times 2z^2$

e) Simplify $3b + \dfrac{b}{2}$

a) There are 3 boxes of n grams and 2 boxes of m grams, so that is $3 \times n$ and $2 \times m$ Writing this correctly as algebra gives $3n + 2m$, or $2m + 3n$. The two expressions are $(3n + 2m)$ grams $(2m + 3n)$ grams	Writing out the masses of the boxes you get $n + n + n + m + m$ $\quad 3n \quad + \quad 2m$ As it is addition, you could swap the order. You could also write: $m + m + n + n + n$ $\quad 2m \quad + \quad 3n$ You can then write the answers using brackets to show the whole amount is measured in grams. $(3n + 2m)$ grams $(2m + 3n)$ grams	$\boxed{n}\ \boxed{n}\ \boxed{n}\ \boxed{m}\ \boxed{m}$ $\boxed{\ \ 3n\ \ }\ \boxed{\ \ 2m\ \ }$ $\boxed{\quad 3n + 2m \quad}$ **Tip** Remember that when your answer is one of the variables you just write the letter. For example, if you have one g you write g and **not** $1g$.
b) The perimeter of the triangle is the sum of the sides, which is: $2c + 2c + d$ This simplifies to $4c + d$ so the perimeter is $(4c + d)$ cm	Write down the terms for the sides, then add together the terms in c. $2c + 2c + d$ $\quad 4c \ \ + d$ The two terms have different letters, so you can't mix them together. Remember to use brackets when you write your answer to show that the whole perimeter is measured in centimetres.	$\boxed{\ 2c\ \ }\boxed{\ 2c\ \ }\boxed{d}$ $\boxed{\quad 4c \quad}\boxed{d}$ $\boxed{\quad 4c + d \quad}$
c) $m + n + 3m + 2n - 2m$ $= m + 3m - 2m + n + 2n$ $= 2m + 3n$	Rearrange the expression so that you have all the terms in m followed by all the terms in n. $m + 3m - 2m + n + 2n$ Remember that the sign in front of the term tells you whether you are adding or subtracting that term. Simplify the m terms, then the n terms.	$m\ +\ n\ +\ 3m\ +\ 2n\ -\ 2m$ $m\ +\ 3m\ -\ 2m\ +\ n\ +\ 2n$ $\qquad\ 2m \qquad + \qquad 3n$
d) $4 \times 2z^2 = 8z^2$	$2z^2$ is the same as $2 \times z^2$ $4 \times 2 \times z^2 = 8z^2$	$\boxed{2z^2}\boxed{2z^2}\boxed{2z^2}\boxed{2z^2}$ $\boxed{z^2}\boxed{z^2}\boxed{z^2}\boxed{z^2}\boxed{z^2}\boxed{z^2}\boxed{z^2}\boxed{z^2}$ $\boxed{\qquad 8z^2 \qquad}$
e) $3b + \dfrac{b}{2}$ $= \dfrac{6b+b}{2} = \dfrac{7b}{2} = \dfrac{7}{2}b$	In 3 whole bs there are 6 half bs. 6 half bs plus another half b makes 7 half bs.	$\boxed{\ b\ }\boxed{\ b\ }\boxed{\ b\ }\boxed{\tfrac{b}{2}}$ $\boxed{\tfrac{b}{2}}\boxed{\tfrac{b}{2}}\boxed{\tfrac{b}{2}}\boxed{\tfrac{b}{2}}\boxed{\tfrac{b}{2}}\boxed{\tfrac{b}{2}}\boxed{\tfrac{b}{2}}$ $\boxed{\qquad \tfrac{7b}{2} \qquad}$

1 Simplify by collecting like terms:

a) $m + 3m + m$ b) $2b + 5b + b$ c) $y + 3y + 2y + y$ d) $4g + 7g + 5g$

e) $5e - 3e$ f) $6m - 5m$ g) $15k - 6k$ h) $5d - d$

i) $9a + a - 2a$ j) $3p - 4p + 5p$ k) $6h + 5h - 10h$ l) $2x + x - 3x$

2 Form an expression for the length, in centimetres, of the perimeter of each triangle. Simplify the expression by collecting like terms.

a)

b)
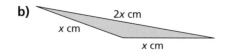

c)
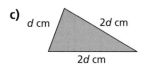

d)
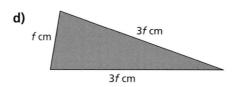

3 Simplify by collecting like terms:

a) $a + b + a$ b) $2x + y + y$ c) $m + n + n + m$ d) $3d + e - e$

e) $3y + 2x + 2y$ f) $3f - f + 4g + 2g$ g) $k + 4c + 2k - k$ h) $3v + 2w + 5w - v$

4 Form an expression, in centimetres, for the perimeter of each shape and simplify by collecting like terms.

a)
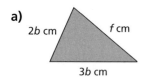

b)
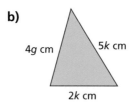

c)
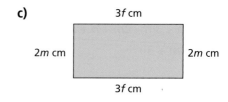

d)
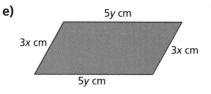

e)

f)
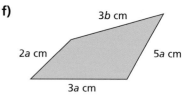

5 Simplify:

a) $5n + 4 - 3n + 1 + 2n$

b) $4a - 3b + 6a - 5b$

c) $7u - 2 + u + 11 - 3u$

d) $8t + 2 - 5t - 12$

e) $e + 4f + 3e - 2f - 6f$

f) $6g - 3h - 2g + 4h + 2$

g) $9 + 4u - 4 - 7u + 3$

h) $3 - 2r + q + 4 - 5r - 7q$

i) $8c - 13 + 4d - 6c + d - 2 - 2c$

j) $6a - 4b + 7c - b - 2c - 11a$

k) $9a^2 - 3a + a^2 + 2a + 3$

l) $4r - 9 - 6r^2 + 4 + 2r^2 - r$

6 Simplify:

a) $3 \times a$

b) $4 \times 2c$

c) $7 \times d \times 4$

d) $c \times c$

e) $3 \times c \times c$

f) $c \times d \times c$

g) $c \div 2$

h) $c \div d$

7 Match each expression in the first column with one from the second column.

$a^2 + 2a^2 + b + b$
$a^2 + 3a^2 - 2a^2$
$2ab - a + 3ab + 2b$
$3 \times 5a^2$
$5a^2 + 3 - a^2 + 7$
$4a^2 + 3b^2 - 3a^2 + b^2$
$ab + 3a + 7ab + 2b$

$2a^2$
$5ab - a + 2b$
$4a^2 + 10$
$3a^2 + 2b$
$8ab + 3a + 2b$
$15a^2$
$a^2 + 4b^2$

> **Tip**
>
> Remember a^2 means '$a \times a$' and is one term of an expression. a^2 and $2a^2$ are like terms, so $a^2 + 2a^2 = 3a^2$.

8 Copy and complete each statement.

a) $6h + \dots\dots + \dots\dots + 7 = 8h + 11$

b) $2h - \dots\dots + \dots\dots + 4k = 5h + k$

c) $7c + \dots\dots - \dots\dots + 11 = 15 - 3c$

d) $5y - 4 - \dots\dots + \dots\dots = 1 - y$

9 a) Write down two expressions that simplify to $8d$.

b) Write down two expressions that simplify to $5a + 3b$.

c) Georgie says that $4y + 3y + 2$ simplifies to $9y$. Is she correct? Give a reason for your answer.

> **Think about**
>
> Can you find expressions to put in the gaps to make this statement correct?
>
> $\dots\dots + \dots\dots = \dfrac{5y}{4}$
>
> Can you complete the statement in more than one way?

10 Write these as a single fraction:

a) $\dfrac{x}{2} + \dfrac{3x}{2}$

b) $\dfrac{a}{4} + \dfrac{2a}{4}$

c) $\dfrac{4b}{5} - \dfrac{2b}{5}$

d) $2a + \dfrac{1}{3}a$

e) $2x - \dfrac{1}{4}x$

f) $\dfrac{x}{2} + \dfrac{x}{4}$

g) $\dfrac{2y}{3} + \dfrac{y}{6}$

h) $\dfrac{2b}{5} - \dfrac{b}{4}$

i) $\dfrac{y}{2} + \dfrac{y}{3}$

j) $\dfrac{3x}{5} - \dfrac{x}{6}$

Thinking and working mathematically activity

In a number pyramid, the number in each box is the sum of the two numbers in the boxes below it.

Copy and complete these number and letter pyramids.

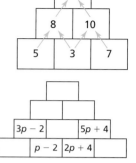

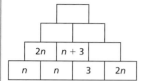
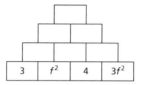
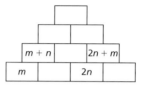

Now create your own number and letter pyramid.

Which blocks could you leave empty so that it is still possible to solve your pyramid?

Challenge a partner with your pyramid.

10.2 Multiplying a constant over a bracket

Key terms

Multiplying a constant over a bracket means removing the brackets from expressions.

For example, $3(6 + w)$ means 3 lots of 6 plus 3 lots of w

$6 + w + 6 + w + 6 + w = 18 + 3w$

This tells you that

$3(6 + w) = 18 + 3w$

If you look at this you can see that both the 6 and the w have been multiplied by 3 so everything inside the brackets has been multiplied by the number outside the brackets.

In the same way, $4(r + 8) = 4r + 32$

and $5(7 - t) = 35 - 5t$

You might be asked to **expand** brackets, and this means to multiply the constant over the brackets like in these examples.

> **Tip**
>
> Make sure you multiply every term inside the brackets by the number outside the brackets.

Worked example 2

Jack spends $2m$ minutes on his homework.

Peter spends 3 minutes more on his homework than Jack.

Daniel spends twice as much time on his homework as Peter.

Form expressions for the amount of time, in minutes, that **a)** Peter and **b)** Daniel spend on their homework.

a) Peter spends $2m + 3$ minutes on his homework.	If Jack spends $2m$ minutes on homework and Peter spends 3 minutes more, then Peter spends $(2m + 3)$ minutes.	$2m$ $\quad$ 3 $2m + 3$

b) Daniel spends $2(2m + 3)$ minutes $2(2m + 3) = 4m + 6$	Daniel spends twice as much time as Peter which is 2 lots of $2m + 3$ You write this as $2(2m + 3)$ Multiply out $2(2m + 3) = 4m + 6$ Every term in the brackets is multiplied by the number outside the brackets.	$2m + 3$ $2m + 3$ $2m$ 3 $2m$ 3 $2m$ $2m$ 3 3 $4m$ 6

Exercise 2

1 Expand these expressions:

   a) $2(a + 1)$ **b)** $2(d + 3)$ **c)** $3(c - 2)$ **d)** $5(h - 5)$

   e) $4(2b + 1)$ **f)** $5(3x - 1)$ **g)** $4(3a + 2b)$ **h)** $3(3d - 2c)$

   i) $4(2a + b + 3c)$ **j)** $6(x - 3y + 4z)$ **k)** $4(3p - q + 2r)$ **l)** $10(5r + 3s - 10t)$

2 Copy these statements. Write whole numbers in the spaces to make them true.

   a) $8w + 16 = $ …….. (…….$w + $ ………)

   b) $42 - 12t = $ ………(…….. $-$ ……..t)

3 Some of these expressions are equal to each other. Sort the expressions into three groups.

   $15(n + 2)$ $9(2n + 3)$ $10n + 40$ $3(5n + 10)$ $10(n + 4)$

     $2(5n + 20)$ $15n + 30$ $18n + 27$ $3(6n + 9)$

4 Ifrah has k stamps in her collection. Rashid has 10 more stamps than Ifrah.

Andrew has double the number of stamps that Rashid has.

   a) Form an expression, in terms of k, for the number of stamps that Rashid has.

   b) Form an expression, in terms of k, for the number of stamps that Andrew has.

5 Saif has a collection of p comics in her collection. Sophie has 4 more comics than Saif. Jana has twice as many comics as Sophie. Deema has three times as many comics as Jana.

   a) Form an expression, in terms of p, for the number of comics Sophie has.

   b) Form an expression, in terms of p, for the number of comics Deema has.

   c) Form an expression for the total number of comics of all four girls.

6 Jane works out the perimeter of this rectangle. Here is her working.

     $2a + 1 + a + 2 = 3a + 3$

     $2(3a + 3) = 6a + 3$

     So the perimeter is $3(2a + 1)$ cm.

Is Jane correct? Explain your answer.

$(a + 2)$ cm

$(2a + 1)$ cm

Thinking and working mathematically activity

With a partner, create a set of 12 matching cards with expressions containing brackets and cards with the same expression multiplied out. Place your cards face down on the table and shuffle them. Take it in turns to choose a pair of cards. If the expressions match, you keep that pair. If the expressions don't match, you put them back. The winner is the first person to get 7 matching pairs.

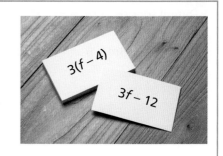

Consolidation exercise

1 Are the following statements true or false?

a) $4c$ means $4 \times c$

b) $t + t + t + t + t + t = 6t$

c) $d + d - d - d = d$

d) $c \times d = d \times c$

e) $a + b = ab$

f) $e + f + 2e + 3f = 2e + 2f$

2 Simplify these expressions:

a) $3a + 4a$

b) $5a + 7a - 8a$

c) $4m + n + m + n$

d) $2k + 7j - 2k + j$

e) $4e - 5f - 3e + 6f$

f) $3x^2 + 7y - 2x^2$

g) $m^2 + 2n^2 + 3m^2 - n^2$

h) $5pq + 3q + 2q - 2pq$

3 Write each of these as a single fraction:

a) $\frac{2m}{3} + \frac{5m}{3}$

b) $\frac{3b}{8} + \frac{b}{4}$

c) $2d + \frac{2}{3}d$

4 Expand:

a) $4(s + 5)$

b) $7(2t - 4)$

c) $3(2d + 5e)$

d) $6(2m - 3n)$

e) $5(p + 2q - 4)$

f) $3(5r - 7s + t)$

5 a) Find an expression for the area of the rectangle. Write your expression with and without brackets.

b) Find an expression for the perimeter of the rectangle. Write your expression as simply as possible.

6

$4a + 3$

c) Find a rectangle with the same area as the rectangle above. Does your rectangle have the same perimeter as the one above?

End of chapter reflection

You should know that...	You should be able to...	Such as...
When more than one term in the same letter is added or subtracted this is called collecting like terms or simplifying an expression.	Collect like terms.	Simplify by collecting like terms: a) $3a + 4b + 2a - 2b$ b) $5x^2 - 2x^2$
When a constant is multiplied over brackets, every term inside the brackets is multiplied by the number outside the brackets. This is often called expanding or multiplying out the bracket.	Multiply a constant over a set of brackets.	Expand: $3(2g + 4)$

11 Angles

You will learn how to:
- Know that the sum of the angles around a point is 360°, and use this to calculate missing angles.
- Derive the property that the sum of the angles in a quadrilateral is 360°, and use this to calculate missing angles.
- Recognise the properties of angles on:
  - parallel lines and transversals
  - perpendicular lines
  - intersecting lines.

Starting point

Do you remember…

- that the sum of the angles on a straight line and in a triangle is 180°?
  For example, find the value of a and the value of b in these triangles.

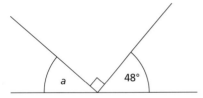
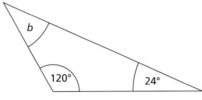

- what parallel and perpendicular lines are?
  For example, write down the name of a side that is parallel to the red line and the name of a side that is perpendicular to the red line.

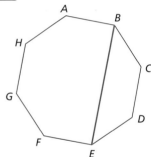

- what is meant by an acute angle, an obtuse angle and a right angle?
  For example, what type of angle is an angle measuring 108°?

- what rotational symmetry is?
  For example, what is the order of rotational symmetry of a parallelogram?

This will also be helpful when…

- you learn more about solving geometric problems.

11.0 Getting started

Mae is investigating angles around a point.

She knows that four right angles join together around a point.

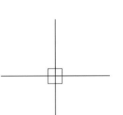

She knows a right angle is 90° and so four of them will add up to 360°.

Mae wants to know if three angles around a point will also add up to 360°.

Draw some diagrams of three angles around a point, measure them and see if they too add up to 360°.

See if four angles around a point will also add up to 360°.

What about more than four angles around a point?

11.1 Angles around a point

Worked example 1

Find the size of angle b. Explain your answer.

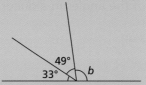

$180 - 33 - 49 = 98$ So $b = 98°$	You need to work out the size of the missing angle. You can do this by taking the two angles that you do know away from 180°.
The three angles are on a straight line. Angles on a straight line add up to 180°.	You are asked to explain your answer. You need to state the angle fact that you have used.

Key terms

The **angles around a point** add up to 360°.

Exercise 1

The diagrams in this exercise are not always drawn to scale.

1 Find the size of each labelled angle.

a)

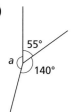

b)

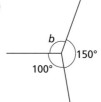

c)

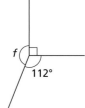

d)

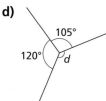

e)

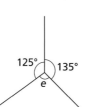

f)

2 Find the size of each labelled angle.

a)

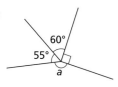

b)

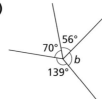

c)

d)

e)

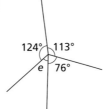

f)

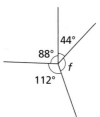

3 Find the size of each labelled angle.

a)

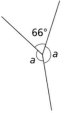

b)

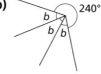

c)

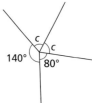

d)

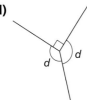

e)

f)

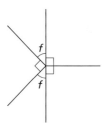

4 Find the value of angle x in each diagram.

a)

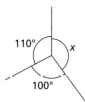

b)

c)

d)

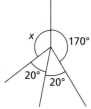

e)

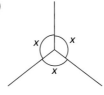

f)

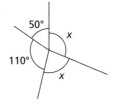

Put each diagram in the correct column of the table by finding each value of x.

The first one has been done for you.

$x = 100°$	$x = 120°$	$x = 150°$	$x = 160°$
		A	

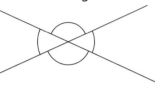

Thinking and working mathematically activity

Draw a diagram showing two straight lines that intersect.

Use a protractor to measure the four angles. What do you notice about the sizes of the angles?
Draw other diagrams with two lines that intersect. Are the angles related in the same way?
Try to explain why the angles are related in this way.

5 Find the size of each labelled angle.

a)

b)
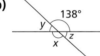

c)
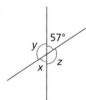

6 Find the size of each labelled angle.

a)

b)

c)

d)
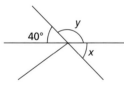

e)
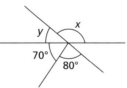

f)
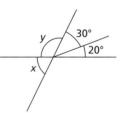

7 Lines AB and CD are perpendicular.

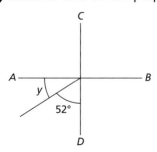

Calculate the value of y.

8 Andrew draws a diagram where three angles are all about the same point.

He says that:

- two of the angles are the same size
- one angle is twice as big as each of the others.

Describe his diagram.

9 Sophia draws two lines that cross. She measures the angles.

She says one of the angles is twice the size of another. What size are the angles?

10 a) Which angle is equal to angle a?

b) What is angle a + angle b?

c) What angle is equal to angle b?

d) What is the sum of angles a, b, c and d?

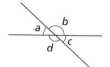

11.2 Angles in a quadrilateral

▼ Thinking and working mathematically activity

Draw a quadrilateral on a piece of paper. Mark the four angles a, b, c and d.

- Carefully cut out the quadrilateral.
- Tear the quadrilateral to separate the four angles as shown.
- Try to rearrange the four angles a, b, c and d to fit together around a point.
- What does this tell you about your four angles?
- Investigate this with other quadrilaterals. Comment on your results.

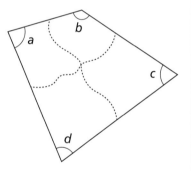

Think about

If you know three angles in a quadrilateral, can you always work out the fourth angle?

If you know fewer than three angles in a quadrilateral, is it possible to work out the other missing angles? What properties might help you to do this?

Worked example 2

Find the size of angle *e*. Give a reason for your answer.

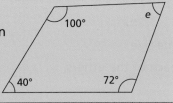

The **angles in a quadrilateral** add up to 360°.

$360 - 40 - 72 - 100 = 148$ So $e = 148°$	The shape is a quadrilateral. You know that the angles will add up to 360°. You need to take the angles that you do know away from 360°.
Angles in a quadrilateral add to 360°.	You are asked to explain your answer. You need to state the angle fact that you have used.

Worked example 3

a) Find the size of angles *a* and *b* in this parallelogram.

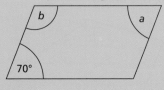

b) Find the size of angle *c* in this kite.

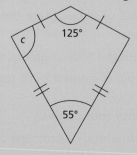

a) $a = 70°$	Angle *a* is opposite the angle 70°, so is equal to it.	Parallelograms have rotational symmetry of order 2. So opposite angles are equal.
$360 - 70 - 70 = 220$ $b = 220 \div 2 = 110°$	All four angles in the parallelogram add up to 360°. The two acute angles add up to 140°. The two obtuse angles are equal and together they add up to $360 - 140 = 220°$	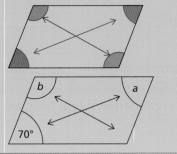

b) $360 - 125 - 55 = 180$ $c = 180 \div 2 = 90°$	The four angles in a kite add up to 360°. The two missing angles must add up to $360 - 125 - 55 = 180°$. So angle c must be $180 \div 2 = 90°$.	A kite has a line of symmetry, so two angles are equal. 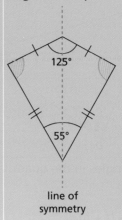

Exercise 2

The diagrams in this exercise have not been drawn to scale.

1 Find the size of each labelled angle.

a)

b)
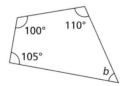

c)
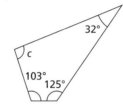

d)
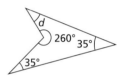

Discuss

Kites have one line of symmetry. Draw a different type of quadrilateral that has one line of symmetry.

2 Are these statements true or false? Explain your answers.

a) A quadrilateral must have at least one acute angle.

b) A quadrilateral cannot have four acute angles.

c) A quadrilateral can have two obtuse angles.

d) A quadrilateral can have three obtuse angles.

3 Put these statements in order to produce a proof that the sum of angles in a quadrilateral is 360°

Look at triangle *BCD*.

Since this is a triangle, $c + g + h = 180°$ because the angles in a triangle sum to 180°.

Divide the quadrilateral into two pieces using a diagonal from one vertex to the opposite vertex.

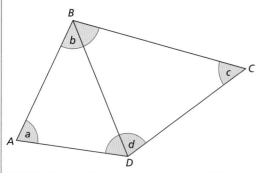

The sum of the angles in the quadrilateral:

$= a + b + c + d$

$= a + (e + g) + c + (f + h)$

$= a + e + f + c + g + h$

$= 180° + 180°$

$= 360°$

Here is a quadrilateral $ABCD$.

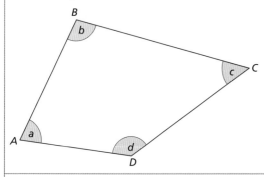

Look at triangle ABD.

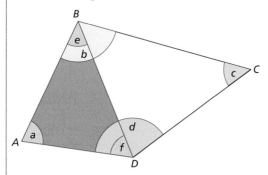

Since this is a triangle, $a + e + f = 180°$ because the angles in a triangle sum to $180°$.

4 Calculate the missing angles in each diagram.

a)

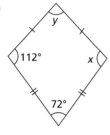

b)

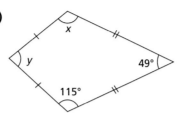

c)

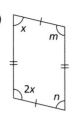

5 Find the size of each labelled angle.

a)

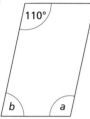

b)

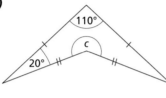

c)

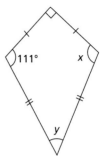

6 Find the size of each labelled angle.

a)

b)

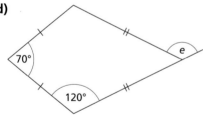

c)

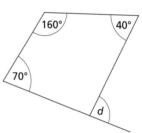

d)

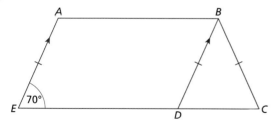

e)

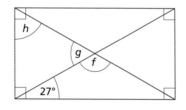

7 *ABDE* is a parallelogram and *BCD* is an isosceles triangle. Find the value of angle *DBC*.

8 *ABDE* is a kite and *BCD* is an isosceles triangle. Calculate the size of angle *EAB*.

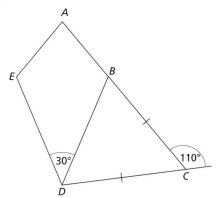

9 Write an angle question for your partner to solve. Draw a diagram with a missing angle, *x*. The answer should be 48°.

10 Here is a diagram of a quadrilateral.

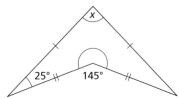

Theo says that angle *x* is 165°.

a) Theo is incorrect. How can you tell this without doing any calculations?

b) What calculation do you think Theo did to get his answer?

c) Find the correct value of *x*.

11 a) Draw a quadrilateral. Draw a diagonal in the quadrilateral joining two vertices.

b) What two shapes has the quadrilateral been divided into?

c) Explain how this shows that the four angles of any quadrilateral must add up to 360°.

11.3 Angles and parallel lines

Key terms

Parallel lines are lines that will never cross however long they are.

A **pair of parallel lines** are indicated by arrows drawn on them.

A **transversal** is a line crossing two or more lines. When a transversal crosses parallel lines, it creates equal angles.

The pairs of angles shaded green in each of these diagrams are equal.

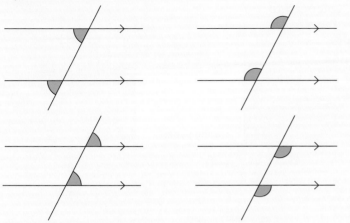

The pairs of angles shaded blue in these diagrams are also equal.

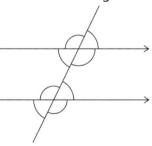

Discuss

If you have a pair of parallel lines and a transversal, how many angles do you need to know to be able to work out all of the angles formed?

Thinking and working mathematically activity

Draw a diagram showing two parallel lines and a transversal.

Measure the size of the angles. What do you notice?

Try with other drawings of parallel lines and a transversal. Do the same relationships always hold?

Worked example 4

In this diagram, which angle is equal to angle a?

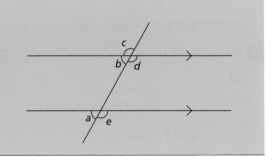

Measure the angles in the diagram using your protractor.	To identify the angle that is equal to angle a, you need to compare the size of the angles.	Use tracing paper to draw around angle a.
$a = 60°$ $b = 60°$ $c = 120°$ $d = 120°$ $e = 120°$		tracing paper
Comparing the sizes of the angles you have measured, you can see that angle a and angle b are the same size.	Compare the size of the angles.	Compare angle a to the other angles using the tracing paper. Angle b is the only labelled angle that matches the tracing of angle a.

Exercise 3

The diagrams in questions 2 to 6 have not been drawn to scale.

1

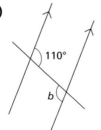

a) Write down all the angles equal to angle a.   b) Write down all the angles equal to angle b.

2 Write down the size of the lettered angles.

a)

b)

c)

d)

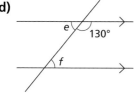

e)

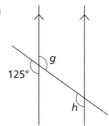

3 Work out the size of the lettered angle in each diagram.

a)

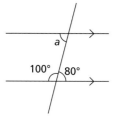

b)

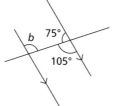

c)

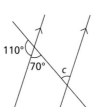

d)

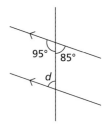

4 Work out the size of the lettered angle in each diagram.

a)

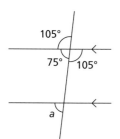

b)

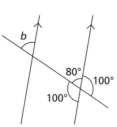

c)

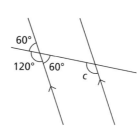

d)

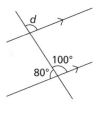

5 Write down pairs of angles that are equal in these diagrams.

a)

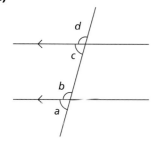

b)

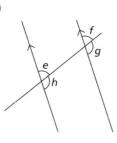

6 Work out the size of the lettered angle in each diagram.

a)

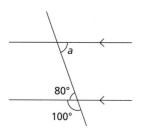

b)

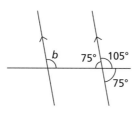

c)

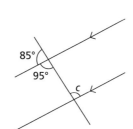

d)

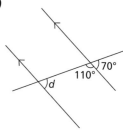

7 Safia has been investigating angles formed by parallel lines and transversals.

She has drawn this diagram.

Safia says that all of the angles formed by a line crossing a pair of parallel lines are equal.

Is Safia correct? Explain your answer.

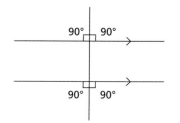

8 **Vocabulary question** Using the diagram, copy and complete the sentences using words from the box.

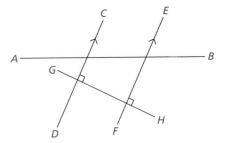

transversal	parallel	perpendicular

CD and EF are _____ .

CD and GH are _____ .

AB is a _____ .

Consolidation exercise

The angles in this exercise have not been drawn to scale.

1 Find the size of each labelled angle.

a)

b)

c)

d)
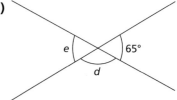

2 Find the size of each labelled angle.

a)

b)

c)

3 Write down whether these statements are true or false.

a) A parallelogram has two pairs of equal angles.

b) A kite has two pairs of equal angles.

c) The angles in a trapezium add up to 360°.

d) A quadrilateral could have two angles of 110° and two angles of 80°.

4 Calculate the size of the missing angle in each quadrilateral.

a)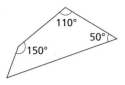

b)

c)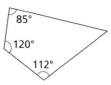

d)

e)

5 Calculate the size of the lettered angles.

a)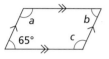

b)

c)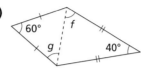

d)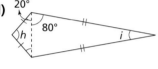

e)

6 Calculate the size of all the lettered angles.

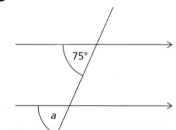

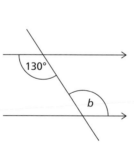

7 The diagram shows a pair of parallel lines and a transversal.

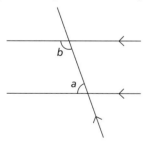

Moses says that angles a and b must add up to 180°.

Investigate whether or not Moses is correct.

End of chapter reflection

You should know that...	You should be able to...	Such as...
The sum of the angles around a point is 360°.	Calculate missing angles around a point.	Find angle x: 110° 80° x
Intersecting lines create a pair of equal angles.	Name the angles that are equal when two lines intersect.	48° y x Write down the size of the angles marked x and y.
The sum of the angles inside a quadrilateral is 360°.	• Demonstrate that the angles of a quadrilateral add up to 360°. • Calculate missing angles in a quadrilateral.	x 45° 117° 77° Calculate angle the angle marked x.
Transversals that cross a pair of parallel lines generate equal angles.	• Recognise equal angles when a transversal is drawn through two parallel lines.	65° 115° x Find the angle marked x.

12 Measures of average and spread

You will learn how to:

- Use knowledge of mode, median, mean and range to describe and summarise large data sets.
- Interpret data, identifying patterns, within and between data sets, to answer statistical questions. Discuss conclusions, considering the sources of variation, including sampling, and check predictions.

Starting point

Do you remember…

- how to find the mode, median, mean and range from a list of results?
- For example, find the mode, median, mean and range of these long jump results:
  5.4 m, 4.8 m, 5 m, 4.6 m, 4.8 m, 3.7 m, 4.8 m, 4.5 m
- how to use a frequency table?

  For example, 27 students were asked how many siblings they have. The results are shown in the frequency table below.

  How many students have fewer than 2 siblings?

  What percentage of the students have no siblings?

Number of siblings	Frequency
0	4
1	7
2	8
3	5
4	3

This will also be helpful when…

- you learn how to compare two data sets using averages and spreads.

12.0 Getting started

Ahmet is looking at two ice skaters' scores in a competition.

	Judge 1	Judge 2	Judge 3	Judge 4	Judge 5	Judge 6	Judge 7	Judge 8
Skater 1	101	99	107	89	121	115	105	103
Skater 2	96	102	113	109	108	105	105	102

- Just by having a quick look at the data, which skater do you think did better?
- How could you check your answer?
- What is the total score for each skater?

- Skater 3 was awarded a total score of 784. Each of the eight judges gave the same score. What was that score?
- Find the highest and lowest scores for each skater and work out how far apart they are.
- Which skater has a bigger spread of scores?

You have just found an average and a measure of spread for this data!

12.1 Mean, median, mode and range

Key terms

The mean is an average.

The **mean** is the total value of all the data divided by the number of pieces of data.

The mode is an average. The **mode (**or the **modal value)** is the value in a set of data that occurs most frequently.

A set of data can have more than one mode (or no mode at all). You can also find the mode for a set of data that are not numerical.

The median is an average.

The **median** is the middle value of data, once the data is arranged in order of size.

When there are n values in a data set, you can use the formula $\frac{n+1}{2}$ to find the position of the median.

If $\frac{n+1}{2}$ is not a whole number, the median will be halfway between the middle pair of values.

The range is a measure of spread.

The **range** is the difference between the highest data value and the lowest data value.

It shows how consistent the data is.

> **Did you know?**
>
> In many sports the awarding of points is done by finding averages of individual judges' scores. Computer programmes usually calculate these instantly when the judges have entered their scores.

Worked example 1

Here is a list of the weights of birds in a wildlife survey, in grams:

131, 128, 104, 154, 176, 129, 122, 136, 197, 144, 101, 98, 122, 165, 123, 108

Find the median for these data.

98, 101, 104, 108, 122, 122, 123, 128, 129, 131, 136, 144, 154, 165, 176, 197	Put the numbers in ascending order.	

There are 16 pieces of data.	Count how many pieces of data there are. This is n in the formula $\frac{n+1}{2}$.	98 … 197, 101 … 176, 104 … 165, 108 … 154, 122 … 144, 122 … 136, 123 … 131, (128 129) ← median
$\frac{16+1}{2} = 8.5$	The median is halfway between the 8th and the 9th pieces of data.	
98, 101, 104, 108, 122, 122, 123, 128, 129, 131, 136, 144, 154, 165, 176, 197 The median is $\frac{128+129}{2} = 128.5$ g	Finding the mean of 128 and 129 is the best way to find the value halfway between them.	128 —— 128.5 —— 129 ↑ median

Worked example 2

Jake has six old coins in his collection. The mean average diameter of his coins is 3.5 cm. The diameters of five of his coins are 2.4 cm, 4.9 cm, 3.1 cm, 5.1 cm and 2.9 cm. Find the diameter of the missing coin.

$3.5 \times 6 = 21$ cm $2.4 + 4.9 + 3.1 + 5.1 + 2.9 = 18.4$ $21 - 18.4 = 2.6$ The missing coin has a diameter of 2.6 cm.	Find the total diameter of all six coins by multiplying the mean by 6. Subtract the total diameter of the other five coins from the total you have just found.	2.4 | 4.9 | 3.1 | 5.1 | 2.9 | ? 21 3.5 | 3.5 | 3.5 | 3.5 | 3.5 | 3.5

Exercise 1

1 Find the mode for each of these sets of data.

a) Favourite colour of 12 people:

red	blue	purple	blue	yellow	blue
purple	blue	purple	green	orange	pink

b) Shoe size of ten people:

39	42	45	38	40	42	35	39	47	41

2 Find the median and range of each set of data.

a) Age of 11 children at a playgroup:

| 3 | 4 | 2 | 3 | 3 | 3 | 2 | 1 | 4 | 0 | 4 |

b) Masses of nine people:

57 kg   64 kg   59 kg   61 kg   68 kg
61 kg   56 kg   67 kg   63 kg

> **Tip**
>
> Remember to include units in your answer.

3 James measured the rainfall in his garden every year over a two-week period in August.
Here are his results to the nearest tenth of a centimetre.

1.5 cm	3.9 cm	0.0 cm	0.7 cm	0.0 cm
5.9 cm	2.4 cm	3.4 cm	4.7 cm	0.0 cm
2.1 cm	4.5 cm	1.7 mm	3.1 cm	

Find:

a) the mode   **b)** the median   **c)** the mean   **d)** the range

4 Jasmine has a set of four number cards.

| 9 | | 11 | | 12 | | ? |

The numbers have a range of 5. Which two possible values could the last card have?

5 Here are the numbers of students absent from Class A over the last two weeks.

| 0 | 3 | 2 | 0 | 0 | 1 | 2 | 4 | 3 | 2 |

Paula says, 'On average there were four students absent because the range is $4 - 0 = 4$'.

Do you agree with Paula? Explain your answer.

6 Here are the number of cars that pass a house each day over 23 days.

10, 22, 28, 24, 32, 14, 20, 26, 19, 24, 20,
6, 8, 15, 28, 16, 12, 20, 32, 15, 19, 22, 5

Kais says the median is 12.

> **Think about**
>
> What strategy will you use to help you to put the data in order efficiently? What can you do to help you check that you have all the data in your list?

a) Explain the error that Kais has made.

b) Find the correct value of the median.

c) Calculate the mean number of cars.

7 A set of five numbers has a mean of 7. Four of the numbers are 4, 5, 11 and 10.
What is the final number?

8 Vocabulary question Match each of the these key terms to its description.

Median	A single value to represent the typical value of data in a set
Mean	The middle value of the data when it is arranged in order
Mode	The difference between the highest and lowest data values
Range	The most common data value

9 The mean height of five people is 1.76 m. A new person joins them. The mean height is now 1.74 m.

Find the height of the new person.

> **Tip**
>
> Think about how the height of the new person affects the total of the heights of all of the people.

10 Real data question The table shows the percentage of the population in 15 countries that live in towns.

Algeria 72.6%	Cameroon 56.4%	Lithuania 67.7%
Angola 65.5%	Denmark 87.9%	Malawi 16.9%
Argentina 91.9%	Fiji 56.2%	Philippines 46.9%
Bahrain 89.3%	Iceland 93.8%	Spain 80.3%
Benin 47.3%	Kazakhstan 57.4%	Turkey 75.1%

Source: UNdata

Enter the data into a spreadsheet. Use a formula to calculate the mean of the percentages.

Thinking and working mathematically activity

Kim has seven numbered cards as shown below.

| 13 | | 12 | | 17 | | 10 | | 18 | | 15 | | ? |

One of her numbers is unknown.

Decide whether or not the missing number can be found if:

- the median of the seven numbers is 14
- the median of the seven numbers is 16
- the median of the seven numbers is 13 and the range is 9

If the missing number cannot be found for certain in any of these, is there anything that can be said about the missing number? Give reasons for your answer.

Worked example 3

Julian is investigating the number of sweets in boxes of fruit drops.
Here are his results:

Number of sweets in a box	Frequency
19	1
20	1
21	4
22	8
23	3
24	3

a) Find the mode.

b) Calculate the mean.

c) Find the range.

d) Find the median.

a) Mode = 22

The mode is the data value that occurs most often, which is 22 since it occurs 8 times.

Number of sweets in a box	Frequency
19	1
20	1
21	4
22	8
23	3
24	3

b) Mean

Number of sweets in a box × Frequency				
19	×	1	=	19
20	×	1	=	20
21	×	4	=	84
22	×	8	=	176
23	×	3	=	69
24	×	3	=	72
		20		440

Mean = $\frac{440}{20}$ = 22

You need to find the total number of sweets and divide by the number of boxes.

You need to start by finding the total number of sweets.

Total number of sweets =

$19 \times 1 + 20 \times 1 + 21 \times 4$

$+ 22 \times 8 + 23 \times 3 + 24 \times 3$

= 440

Altogether there are 20 boxes of sweets in the sample.

So mean = 440 ÷ 20 = 22

Number of sweets in a box	Frequency
19	1
20	1
21	4
22	8
23	3
24	3

19, 20, 21, 21, 21, 21, 22, 22, 22, 22, 22, 22, 22, 22, 23, 23, 23, 24, 24, 24

Total =

19 + 20 + 21 + 21 + 21 + 21 + 22 + 22 + 22 + 22 + 22 + 22 + 22 + 22 + 23 + 23 + 23 + 24 + 24 + 24

= 440

c) Range = 24 − 19 = 5	The highest number of sweets in a box was 24, the lowest was 19. So the range is the difference between these, which is 5.	Range of 5 19 ← — — — — → 24 lowest value highest value
d) Median 10th value = 22 11th value = 22 The median is halfway between the 10th and 11th values. So the median is 22.	The median is the middle piece of data. Use this table to find its location. The median is halfway between the 10th and 11th pieces of data. The 10th and 11th pieces of data are both 22 so the median is 22.	Number of sweets in a box / Location in the data set: 19 — 1st 20 — 2nd 21 — 3rd to 6th 22 — 7th to 14th 23 — 15th to 17th 24 — 18th to 20th

Think about

In the worked example, the mean, median and mode were all the same value. Do you think this will usually be the case?

Exercise 2

1 Jessie rolled an ordinary dice and recorded the scores in this table.

   a) How many times did Jessie roll the dice?
   b) What was Jessie's modal score?
   c) What was the range of scores?

Score	Frequency
1	7
2	8
3	7
4	9
5	6
6	5

2 Lars completed a survey of the number of hours of Science homework students received each week. His results are shown in the table.

Number of hours	Frequency	Number of hours × frequency
0	1	
1	4	
2	5	
3	6	
4	4	

a) Copy and complete the final column of the table.

b) Find the total number of hours of homework completed by the students.

c) Calculate the number of students in Lars's survey.

d) Calculate the mean number of hours of homework completed by the students.

3 Find the mean and median from these tables.

a)

Value	Frequency	Value × frequency
13	2	
14	1	
15	7	
16	5	
17	5	

b)

Value	Frequency	Value × frequency
0	7	
1	5	
2	1	
3	3	

Discuss

How can you check whether your answer is reasonable when checking the mean? Would it be possible to have data going from a lowest value of 3 to a highest value of 13 and a mean of 20?

4 Fatima keeps a record of the number of portions of fruit and vegetables that she eats per day over a number of days.

Her results are shown in the table.

a) What is the modal number of portions of fruit and vegetables?

b) Calculate the number of portions of fruit and vegetables eaten by Fatima in total.

c) For how many days did Fatima record data?

d) Calculate the mean number of portions of fruit and vegetables eaten by Fatima.

e) Fatima says, 'The range of my data is 12 − 2 = 10.'

Number of portions	Frequency
0	3
1	5
2	8
3	12
4	10
5	10
6	2

Do you agree with Fatima? Explain your answer.

5 A flower shop claims that the mean number of flowers in its bouquets is 12.

Kim buys 20 bouquets from the shop and counts the number of flowers.

Number of flowers	Frequency
9	1
10	2
11	3
12	7
13	3
14	3
15	1

Comment on the shop's claim about the mean number of flowers. Show your working.

6 The table shows the number of potholes on some roads in a town centre.

Number of potholes	5	6	7	8	9
Frequency	5	8	6	3	5

Hasnain is asked to find the mean number of potholes per road.

This is his working out:

$5 + 8 + 6 + 3 + 5 = 27$

Mean $= \frac{27}{5}$

Mean = 5.4 potholes per road.

a) Explain why Hasnain's working out is incorrect.

b) Write down some clear instructions that Hasnain could use to correctly calculate the mean.

7 Rageh copied this table from his exercise book at school.

When he got home he realised that he had copied down one number incorrectly in the frequency column.

Given that the total of 25 and the mean of 1.88 are both correct, find the error.

You must show all your working out.

Number of pets	Frequency
0	4
1	7
2	6
3	3
4	3
5	1
Total	25

> **Tip**
>
> First calculate the total number of pets in the class using the two values which you know are correct.

8 **Technology question** Aidan records the time to run 5 km in his school running club, to the nearest minute. All the runners are from the final year in the school.

Time (minutes)	Frequency	Frequency × Time
30	2	
31	0	
32	6	
33	3	
34	8	
35	7	
36	6	
37	0	
38	5	
39	3	

Enter this data into a spreadsheet and use a formula to complete the final column.

a) Use this extra column to help you to find the mean.

b) Aidan says that he can use this data to make a prediction of the average running time for the whole school.

Explain why Aidan should be cautious about using this data when making predictions for the whole school.

9 Sonja wants to investigate the number of magazines bought last week by the students in her school.

She collects data from a random sample of 40 students.

Number of magazines bought last week	Number of students
0	17
1	16
2	5
3	2

a) Calculate the mean number of magazines bought last week by the students in her sample.

b) Write down an estimate for the mean number magazines bought last week by all the students in her year group.

10 a) Create a set of ten different numbers where the mean is less than the mode.

b) Create a set of ten different numbers where the median is less than the mode.

c) Create a set of ten different numbers where the mode is less than the mean.

Thinking and working mathematically activity

Here is a partly completed frequency table.

Value	Frequency
10	
11	
12	
13	
14	
Total	15

Find possible values for the frequencies to show that the following statements are true.

- The median value must be 12
- The range must be 4
- There must be a modal value
- The mean is never a whole number.

Find a set of frequencies so that:

- the mean, median and mode are all 14.

Consolidation exercise

1 Colin has four numbers.

The mode of the numbers is 1

The median of the numbers is 1.5

The mean of the numbers is 2.5

What are Colin's numbers?

2 a) Here are the overnight temperatures recorded each day in London over a two week period.

6 °C	11 °C	9 °C	8 °C	0 °C	2 °C	1 °C
–1 °C	4 °C	–3 °C	0 °C	1 °C	–2 °C	5 °C

Calculate the mean temperature in London over these two weeks.

b) Here are the temperatures recorded in New York over two weeks in winter one year.

Temperature (°C)	Frequency
1	5
0	4
–1	4
–2	0
–3	1

Calculate the mean temperature in New York for this two week period.

3 Here are the number of cars sold by a car showroom each day over the last few weeks.

Number of cars	Frequency
0	3
1	7
2	4
3	1
4	2
5	1

a) Write the list of data that the table has come from.

b) Find the median number of cars sold.

c) How could you find the median directly from the table?

4 Jonathan wanted to find out how many hours students in his class spent online on a particular Saturday. Here are the results of his survey.

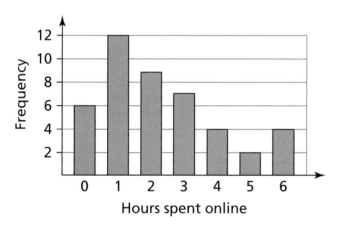

Hours spent online

> **Tip**
>
> Begin by writing the data in a frequency table.

a) Calculate the range.   b) Calculate the mean.

5 This set of five cards has the same mean and median.

Find possible values of A and B.

End of chapter reflection

You should know that...	You should be able to...	Such as...
The mode is the most common piece of data. A set of data can have one mode, more than one mode, or no mode at all.	Find the mode for numerical or categorical data.	Work out the mode of the following colours. blue, red, blue, green, orange red, yellow, black, green, red, yellow, orange, blue
The median is the middle number when the data is put in order.	Arrange n pieces of data in ascending order and find the position of the median using the formula $\frac{(n+1)}{2}$.	Find the median of the following data: 21, 11, 16, 8, 19, 25, 13, 17
The mean is the sum of the data divided by how many pieces of data there are.	Find the mean of data that is displayed in a list or in a frequency table.	Find the mean time.

Time (min)	Frequency
10	3
11	9
12	0
13	2

The range is the highest data value minus the lowest data value.	Identify the highest and lowest values from a list of data, in any order, and use these values to find the range of the data.	Find the range of the following data. 102, 96, 124, 99, 96, 118, 106, 112

13 Calculations

You will learn how to:

- Understand that brackets, positive indices and operations follow a particular order.
- Use knowledge of common factors, laws of arithmetic and order of operations to simplify calculations containing decimals or fractions.

Starting point

Do you remember…

- that the four operations follow a particular order?

  For example, explain why $4 + 2 \times 3 = 10$ and $20 - 10 \div 2 = 15$

- how to use brackets to change the order of operations?

  For example, calculate $3 + 2 \times 6$ and $(3 + 2) \times 6$

- how to use laws of arithmetic and order of operations to simplify integer calculations?

  For example, simplify and work out $4 \times 17 \times 25$ without using a calculator.

This will also be helpful when…

- you learn to use the correct order of operations in calculations that include square roots and cube roots.

13.0 Getting started

| 4 | 4 | 4 | 4 |

How many of the numbers 1 to 20 can you make using:

- four 4s
- the four operations: add, subtract, multiply and divide
- and brackets?

You do not have to use all the 4s each time.

Record your working, making sure your order of calculation is clear.

Which numbers cannot be made? Why not?

- Can you make any numbers greater than 20?
- What is the largest number you can make?
- Can you make a negative number?
- Can you make any decimals or fractions?

  For example, 0.5, 0.25, 0.2, $\frac{1}{3}$

Did you know?

As recently as the 1920s, mathematicians did not all agree about the order of multiplication and division. Some wanted to do multiplications before divisions, while others wanted to do them all in order from left to right (as we do now).

13.1 Order of operations

Key terms

The **order of operations** is the agreed order of calculating in mathematics. The order is:

1. Brackets
2. Indices (powers and roots)
3. Division and Multiplication (from left to right)
4. Addition and Subtraction (from left to right)

Tip

You will learn later that you can *sometimes* rewrite the orders of additions and subtractions, and of divisions and multiplications, to make a calculation easier.

Worked example 1

Calculate

a) $10 + 2 \times 3^2$ b) $(16 - 14 + 1)^3$ c) $(12 \div 2) \times (4^2 + 4)$

a) $10 + 2 \times 3^2$ $= 10 + 2 \times 9$ $= 10 + 18$ $= 28$	Start with the **power**. $3^2 = 9$ Next do the **multiplication**. $2 \times 9 = 18$ Finally do the **addition**. $10 + 18 = 28$
b) $(16 - 14 + 1)^3$ $= 3^3$ $= 27$	Start with the **brackets**. These contain a subtraction and an addition. Work from left to right. $16 - 14 + 1 = 3$ Finally do the **power**. $3^3 = 27$
c) $(12 \div 2) \times (4^2 + 4)$ $= 6 \times 20$ $= 120$	Start with each of the **brackets**. $12 \div 2 = 6$ and $4^2 + 4 = 16 + 4 = 20$ Finally do the **multiplication**. $6 \times 20 = 120$

Exercise 1

1–7

1 Calculate:

a) $12 + 9 \div 3$ b) $15 + 6 - 2$ c) $6 \times 3 + 5$ d) $4 \times 10 \div 5$

e) $24 - 5 \times 2$ f) $18 + 5 - 3$ g) $6 + 9 \div 3 + 7$ h) $6 - 4 \times 4 + 1$

i) $30 - 4^2 + 1$ j) $28 \div 4 \times 3$ k) $4 \times 7 - 2^3$ l) 4×3^2

2 Explain why the calculations $6 - 18 \div 9$ and $6 - (18 \div 9)$ give the same answer.

3 Li Jing is calculating $12 - 8 + 2$.

She says, 'I need to do the addition before the subtraction, so I get $12 - 8 + 2 = 12 - 10 = 2$.'

Do you agree with Li Jing? Explain your answer.

4 Write down the letters of the calculations that have an answer of 40.

A $3 \times 2^2 + 4$

B $10 - 5 \times 2^3$

C $(13 - 2 \times 3)^2 - 9$

D $120 \div (3 \times 11 - 5 \times 6)$

E $(15 - 3)^2 \div 2 + 4$

F $(9 \times 8 - 20) - (21 - 9)$

5 Calculate:

a) $(5 + 5^2) \times 2$

b) $(6 + 4) \times (3 + 8)$

c) $60 + (15 \div 5)^2 \times 2$

d) $6 + 3 \times ((3 - 1) \times 5)$

e) $100 - (10^2 + 3 \times 30)$

f) $18 \div (14 - (3^2 - 4))$

> **Tip**
>
> When there is one set of brackets inside another set of brackets, work out the inner brackets first.

6 Catelyn wants to work out $(2 + 3)^3$. Her working is below.

$(2 + 3)^3 = 2^3 + 3^3 = 8 + 27 = 35$

Explain what she has done wrong.

Find the correct answer.

7 Add one pair of brackets to make each calculation correct.

a) $3 + 5 \times 5 - 2 = 18$

b) $3 \times 2 + 8 \times 3 = 90$

c) $6 - 2^2 \div 8 = 2$

8 Technology question

a) Use a scientific calculator to calculate $18 + 7 \times 6 - 2$.

b) Use a non-scientific calculator to calculate $18 + 7 \times 6 - 2$.

c) Explain the results.

Thinking and working mathematically activity

By writing one set of brackets in different places, how many different answers can you get to this calculation?

$7 + 8 \times 20 - 12 \div 2$

- If you find different ways of writing brackets that give the same answers, group these together and try to explain why they give the same answers.

- How do you know you have found all possible different answers?

> **Think about**
>
> Use a calculator to calculate -3^2 (using the sign key to type the negative sign) and $(- 3)^2$. Why do you get different answers?

13.2 Simplifying calculations with decimals and fractions

Key terms

The **commutative** law tells you that for addition and multiplication, you can change the order of the numbers without changing the result. For example, 3 + 5 = 5 + 3 and 2 × 3 × 7 = 3 × 7 × 2.

To **partition** a number is to break it into two parts that are added or subtracted. For example, you can partition 7 into 2 + 5, 1 + 6, 10 − 3, and so on.

The **distributive** law tells you that to multiply by a number, you can partition the number first and then multiply by each part. For example, 6 × 13 = 6 × 10 + 6 × 3 and 6 × 19 = 6 × 20 − 6 × 1.

> **Did you know?**
>
> The distributive law also works in reverse. For example, you can rewrite 6 × 10 + 6 × 3 as 6 × 13.

Worked example 2

Find ways to make these calculations easier, and find the answers.

a) 5.7 + 1.4 + 8.6

b) 2.5 × 14 × 4

c) 1.2 + 2.9 − 0.2

d) 60 × 1.6 ÷ 6

e) $56 \times \frac{5}{7}$

a) 5.7 + 1.4 + 8.6 = 1.4 + 8.6 + 5.7 = 10 + 5.7 = 15.7	Changing the order of the numbers in an addition does not change the answer. Swap numbers so the calculation starts with 1.4 + 8.6. This gives you 10, which is easier to work with.	Harder method: 5.7 + 1.4 + 8.6 = 7.1 + 8.6 = 15.7
b) 2.5 × 14 × 4 = 2.5 × 4 × 14 = 10 × 14 = 140	Changing the order of the numbers in a multiplication does not change the answer. Rewrite so the calculation starts 2.5 × 4, and then work from left to right. This gives you 10, which is easier to work with.	Harder method: 2.5 × 14 × 4 = 35 × 4 = 140
c) 1.2 + 2.9 − 0.2 = 1.2 − 0.2 + 2.9 = 1 + 2.9 = 3.9	If a calculation contains **only** additions and subtractions, you can rewrite the order of these without changing the answer. Rewrite as 1.2 − 0.2 + 2.9 and then work from left to right.	Harder method: 1.2 + 2.9 − 0.2 = 4.1 − 0.2 = 3.9

d) $60 \times 1.6 \div 6 = 60 \div 6 \times 1.6$ $= 10 \times 1.6$ $= 16$	If a calculation contains **only** multiplications and divisions, you can rewrite the order of these without changing the answer. Rewrite so the calculation starts $60 \div 6$ and then work from left to right.	Harder method: $60 \times 1.6 \div 6 = 96 \div 6$ $= 16$
e) $56 \times \frac{5}{7} = 56 \div 7 \times 5$ $= 8 \times 5$ $= 40$	Other ways of writing this calculation are $56 \times 5 \div 7$, or $56 \div 7 \times 5$. It is easier to work out $56 \div 7$ first.	Harder method: $56 \times \frac{5}{7} = \frac{280}{7} = 40$

> **Tip**
>
> In parts **c)** and **d)** of Worked example 2, you cannot swap the numbers only. For example, the calculation $1.2 + 0.2 - 2.9$ has a different answer from $1.2 + 2.9 - 0.2$. Check this yourself.

Worked example 3

Find the answer to these calculations. Find ways to make the calculation easier.

a) 6.8×9

b) $1.6 \times 7 + 1.6 \times 3$

a) 6.8×9 $= 6.8 \times (10 - 1)$ $= 6.8 \times 10 - 6.8 \times 1$ $= 68 - 6.8$ $= 61.2$	You can partition 9 into $10 - 1$. Nine 6.8s is the same as ten 6.8s minus one 6.8 because multiplication is distributive.	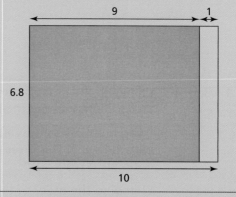
b) $1.6 \times 7 + 1.6 \times 3$ $= 1.6 \times (7 + 3)$ $= 1.6 \times 10$ $= 16$	You can group the seven 1.6s with the three 1.6s to make ten 1.6s. This is because multiplication is distributive.	 

Exercise 2

1 Write down the value of:

a) $247 + 63 - 247$

b) $7.568 + 37.76 - 37.76$

c) $8.7 \times 15 \div 15$

d) $\frac{5}{16} + \frac{8}{21} - \frac{5}{16}$

e) $243 \times 15.7 \div 243$

f) $\frac{9}{11} \times \frac{17}{6} \div \frac{9}{11}$

2 Find:

a) $65 + 134 + 35$

b) $0.25 + 0.91 + 0.75$

c) $\frac{3}{4} + \frac{4}{5} + \frac{1}{4}$

d) $6.7 \times 0.5 \times 2$

e) $\frac{1}{4} \times \frac{3}{8} \times 4$

f) $3\frac{3}{5} + 1\frac{4}{9} + \frac{2}{5}$

3 Match each calculation on the first line with an equivalent calculation on the second line.

> A 5.5×9

> B 5.5×99

> C 5.5×8

> D $5.5 \times 10 - 5.5 \times 2$

> E $5.5 \times (10 - 1)$

> F $5.5 \times 100 - 5.5$

4 In each part, change the calculation to make it easier (without breaking the order of operations rules!) Then find the answer.

a) $178 + 46 - 78$

b) $32 \times 7 \div 8$

c) $12.5 + 3.7 - 0.5$

d) $\frac{8}{9} - \frac{3}{4} + \frac{1}{9}$

e) $4 \times 8.31 \times 2.5$

f) $5 \times 0.45 \times 20$

g) $70 \times 32.7 \div 7$

h) $1\frac{3}{8} + \frac{4}{7} - \frac{3}{8}$

i) $3.7 + 8.6 + 6.3$

j) $\frac{5}{8} \times \frac{3}{10} \div \frac{1}{8}$

k) $3 \times 42 \times \frac{2}{3}$

l) $1.36 + 0.28 - 0.36$

5 Amina says that $4.5 \times 8 \div 3 = 4.5 \times 3 \div 8$

Is Amina correct? Explain your answer.

6 Find the answer to these calculations. Try to find an easy way to work out each one.

a) 4.5×9

b) 3.7×9

c) 5.2×99

d) $2.8 \times 6 + 2.8 \times 4$

e) $1.4 \times 97 + 1.4 \times 3$

f) $\frac{1}{4} \times 33 + \frac{1}{4} \times 7$

7 Rewrite each calculation to make it easier, and then find the answer.

a) 9.9×6

b) 3.25×8

c) $\frac{15}{16} \times \frac{8}{9}$

d) $\frac{9}{32} \times \frac{16}{27}$

Discuss

Chinatsu says, 'I can rewrite the calculation:
$2.8 \times 88 + 2.8 \times 12$
as:
$(2.8 \times 88) + (2.8 \times 12)$
without changing the answer.'

Jonty says, 'Writing the brackets changes the answer.' Discuss who you agree with, and why.

Think about

Nikau says, 'Subtraction is really the same operation as addition, because subtraction is the addition of a negative number. Division is the same operation as multiplication, because division is multiplication by a reciprocal.' Do you agree?

Thinking and working mathematically activity

Explore correct and incorrect ways of reordering calculations by exploring these questions.

Remember to use the order of operations rules when doing the calculations.

- Which of these calculations have the same answer as $20 - 5 - 2$?
  $20 - 2 - 5$ or $5 + 20 - 2$ or $5 - 2 + 20$ or $2 - 20 - 5$

- Which of these calculations have the same answer as $20 + 5 - 2$?
  $5 + 20 - 2$ or $20 - 2 + 5$ or $5 - 2 + 20$ or $20 - 5 + 2$

- Which of these calculations have the same answer as $20 \div 5 \div 2$?
  $20 \div 2 \div 5$ or $5 \div 20 \div 2$ or $2 \div 5 \div 20$ or $5 \div 2 \div 20$

- Which of these calculations have the same answer as $20 \div 5 \times 2$?
  $20 \div 2 \times 5$ or $5 \div 20 \times 2$ or $20 \times 2 \div 5$ or $5 \times 2 \div 20$

- Which of these calculations have the same answer as $20 \times 5 \div 2$?
  $5 \div 2 \times 20$ or $5 \times 20 \div 2$ or $20 \div 2 \times 5$ or $20 \times 2 \div 5$

Describe any pattern you can see.

Consolidation exercise

1 Complete the missing calculation and answer to make five matching pairs.

$6 \times 3 + 2^2$
$6 \times (3 + 2)^2$
$(6 \times 3 + 2)^2$
$(6 \times (3 + 2))^2$

42
22
900
150

2 State whether each calculation is correct or incorrect.

a) $\frac{3}{7} + \frac{4}{5} + \frac{5}{7} = \frac{4}{5}$

b) $\frac{2}{3} \times \frac{8}{9} \times \frac{3}{2} = \frac{8}{9}$

c) $5 \times 6.4 \div 2 = 64$

d) $63 + 5.8 + 27 = 105.8$

e) $3.6 \times 9 + 3.6 = 39.6$

f) $1.7 \times 94 + 1.7 \times 6 = 170$

3 Liam has calculated $24 - 3 + 2 \times 5$ to be 11.

Do you agree with Liam? Explain your answer.

4 Insert brackets in this calculation to make it correct.

$17 - 9 \times 1 + 10 \div 2 = 44$

5 Write a squared symbol in each calculation to make it correct.

a) $5 - 3 + 2 = -2$

b) $2 + 3 \times 5 = 77$

c) $3 \times (1 + 4) = 75$

6 Is each statement always, sometimes or never true? Explain your answers, giving examples.

a) If you change the order of the numbers in a subtraction calculation, it changes the answer.

b) If you change the order of the numbers in a division calculation, it changes the answer.

c) If you do the multiplications and divisions in a calculation in any order, you get the same answer as if you do the multiplications and divisions from left to right.

End of chapter reflection

You should know that...	You should be able to...	Such as...
The correct order of operations in mathematics is: 1. Brackets 2. Indices (powers and roots) 3. Multiplication and Division 4. Addition and Subtraction	Calculate in the correct order.	Calculate: **a)** $(7 - 2)^2 \times 4 + 20 \div 5$ **b)** $(11 - 2 \times 4) \times (6 + 3)$
You can rewrite some calculations with decimals or fractions to make them easier to do, without changing the results.	Simplify calculations by changing the order of numbers or calculations, without breaking the order of operations. Simplify calculations by using the distributive law to partition or group numbers.	Simplify and then work out: **a)** $2.5 \times 1.72 \times 4$ **b)** $\frac{5}{8} \times \frac{7}{13} \times \frac{1}{8}$ **c)** 11×2.3 **d)** $5.7 \times 8 + 5.7 \times 2$

14 Functions and formulae

You will learn how to:

- Understand that a function is a relationship where each input has a single output.
  Generate outputs from a given function and identify inputs from a given output by considering inverse operations (linear and integers).
- Understand that a situation can be represented either in words or as a formula (single operation) and move between the two representations.

Starting point

Do you remember…

- how to use letters to represent unknown numbers?

  For example, the number of apples a person buys could be written as n apples.

- how to write and simplify algebraic expressions?

  For example, writing $a + a + a + 2 = 3a + 2$

- what a term, a coefficient, an expression and an equation are?

  For example, in the equation $4d - 5 = 12$, $4d - 5$ is an expression containing two terms, $4d$ and -5. 4 is the coefficient of d.

- the correct order of operations used to calculate with numbers and algebra?

  For example, in $12 - 2c$, the multiplication is calculated first.

This will also be helpful when…

- you learn to draw graphs and solve equations to find missing numbers.

14.0 Getting started

A mobile phone company charges $12 per month for unlimited texts and calls.

- How much will texts and calls cost you for a whole year?
- You have paid a total of $60 so far for texts and calls. How long have you had the phone?
- Ellis says that the weekly cost for texts and calls is $3. Is he correct?

Discuss your answers with a partner.

Write down two more questions to ask your partner about the cost of texts and calls from this phone company.

14.1 Functions and mappings

Key terms

A **function** is a mathematical process that converts an **input** value (a number you put into the function) to an **output** value (the answer you get out of the function).

You can show a function using a **function machine**, or using algebra as a **mapping** or a **formula**, which shows how two or more variables are connected.

For example, the function 'add 2 to a number' can be written as:

a function machine $input \rightarrow \boxed{add\ 2} \rightarrow output$

a mapping $x \mapsto x + 2$

a formula $y = x + 2$

You can show the effect of a function using a table of input and output values or a mapping diagram.

For example, for $y = x + 2$

input	output
1	3
2	4
6	8
10	12

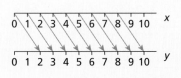

An **inverse function** undoes whatever the original function did. For example, if the original function adds 2, the inverse function will subtract 2.

> **Did you know?**
>
> Function machines are sometimes known as number machines.

Worked example 1

a) Here is a function machine.

$input \longrightarrow \boxed{subtract\ 1} \longrightarrow output$

Complete the table of inputs and outputs for the machine.

input	output
1	
5	4
7	6
10	

b) Another function machine is described as: $x \rightarrow 2x$

Represent this function on a mapping diagram.

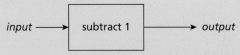

c) Draw the inverse function machine for $y = x - 4$. Find the input of the original function if the output is 15.

a) For an input of 1:

$1 - 1 = 0$

For an input of 10:

$10 - 1 = 9$

input	output
1	0
5	4
7	6
10	9

The function says 'subtract 1' so you subtract 1 from any number that you put into the function machine to get your output.

You can write the operation rather that the words in the function machine.

$1 \rightarrow \boxed{-1} \rightarrow 0$

$10 \rightarrow \boxed{-1} \rightarrow 9$

b)

0 1 2 3 4 5 6 7 8 9 10
0 1 2 3 4 5 6 7 8 9 10

$x \mapsto 2x$ means that you multiply each number that you put in by 2.

You could also write the mapping as a function machine:

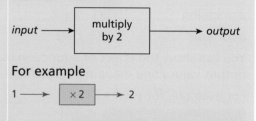

For example

$1 \longrightarrow \boxed{\times 2} \longrightarrow 2$

$2 \longrightarrow \boxed{\times 2} \longrightarrow 4$

$5 \longrightarrow \boxed{\times 2} \longrightarrow 10$

c) $y = x - 4$

Inverse :

$input \leftarrow \boxed{+4} \leftarrow output$

$19 = 4 + 15$

The original function subtracts 4, so has this function machine:

$input \rightarrow \boxed{-4} \rightarrow output$

The inverse undoes the function, so start with the output and go back to the original input.

$y = x - 4$

$input \rightarrow \boxed{-4} \rightarrow output$

$input \leftarrow \boxed{+4} \leftarrow output$

$19 \leftarrow \boxed{+4} \leftarrow 15$

Exercise 1

1 Copy and complete the table of inputs and outputs for these function machines:

a) $input \rightarrow \boxed{\times 5} \rightarrow output$

input	output
1	5
4	20
7	
10	
	25

b) $input \rightarrow \boxed{+4} \rightarrow output$

input	output
1	5
3	
8	
10	14
	17

c) input → | − 7 | → output

input	output
8	
10	3
12	
	17
25	

d) input → | ÷ 3 | → output

input	output
3	1
6	
9	
15	
	10

2 For each table of inputs and outputs, suggest a function machine.

a)

input	output
1	3
2	6
3	9
4	12
5	15

b)

input	output
1	5
2	6
3	7
4	8
5	9

c)

input	output
4	1
8	2
12	3
16	4
20	5

3 a) Copy and complete mapping diagrams for these functions.

You only need to show the mappings that will fit on the number lines.

0 1 2 3 4 5 6 7 8 9 10

0 1 2 3 4 5 6 7 8 9 10

(i) $x \mapsto x + 3$ **(ii)** $x \mapsto 3x$ **(iii)** $x \mapsto x - 2$

(iv) $x \mapsto 2x$ **(v)** $x \mapsto \frac{x}{2}$ **(vi)** $x \mapsto 4x$

b) Look at your diagrams. What do you notice about the gap between the bottom numbers on each diagram and the coefficient of x in the function?

4 What is the inverse function for $x \mapsto x + 5$?

Use the inverse function to copy and complete the table of inputs and outputs for the function $x \mapsto x + 5$

input	output
	5
2	7
	10
	12
	18

5 For each of these functions, draw the function machine and the inverse function machine.

a) $y = x + 10$ **b)** $y = 4x$ **c)** $y = x - 3$ **d)** $y = \frac{x}{3}$

6 Match each function machine with the correct table.

a) $x \mapsto x - 6$ b) $x \mapsto \frac{x}{4}$ c) $x \mapsto 5x$ d) $x \mapsto x + 7$

(i)
input	output
1	5
3	15
4	20
7	35

(ii)
input	output
8	2
9	3
12	6
20	14

(iii)
input	output
1	8
3	10
5	12
9	16

(iv)
input	output
4	1
12	3
24	6
36	9

7 Afia has part of a table for a function machine.

She thinks that the function is $x \mapsto x + 4$

Peter thinks the function is $x \mapsto 3x$

Who is correct? Explain your answer.

input	output
2	6

Thinking and working mathematically activity

The number 3 is used as the input of both these function machines.
The output is the same for both function machines.

3 → [× a] →   3 → [+ b] →

a and b are both positive whole numbers.
Find possible values of a and b.
What do you notice about the possible values of b?
Now investigative with a different input value.
What do you notice about the possible values of b this time?
Can you make any general statements about a and b?

14.2 Constructing and using formulae

Key terms

A **formula** is a mathematical rule written algebraically that shows how two or more variables are connected.

For example, to calculate the number of minutes in any number of hours, you can use the formula $m = 60h$, where m is the number of minutes and h is the number of hours.

A formula doesn't have practical meaning unless you know what the variables represent.

Worked example 2

a) Write in words a formula to change years into months.

b) Write a formula to calculate the number of months, m, in y years.

c) Use your formula to calculate the number of months in 6 years.

a) number of months = number of years × 12	To change years into months you multiply the number of years by 12.	1 year / 12 months
b) The total number of months, m, can be written as $m = y \times 12$ or $12 \times y$ $m = 12y$	To find the number of months in y years, multiply y by 12.	y years: 1 year / 12 months (×5) ... 1 year / 12 months = $12y$ months
c) $m = 12y$ $m = 12 \times 6$ $m = 72$ so there are 72 months in 6 years	You want to find the number of months in 6 years, so $y = 6$. Replace the y in the formula with 6 to find the number of months, m.	6 years: 1 year / 12 months (×6) = 72 months

Exercise 2

1. Using the formula $h = 7t$:

   a) find h when $t = 2$
   b) find h when $t = 7$
   c) find h when $t = 20$

2. Using the formula $c = n - 5$:

   a) find c when $n = 12$
   b) find c when $n = 15$
   c) find c when $n = 4$

3 Kerri uses this formula to find out how many medium-sized potatoes she needs to make one portion of French Fries:

Number of potatoes = number of portions of French Fries × 2

   a) How many potatoes does she need to make 5 portions of French Fries?

   b) How many potatoes does she need to make 20 portions of French Fries?

   c) Write a formula to calculate the number of potatoes, n, needed for f portions of French Fries.

4 Janina charges $20 per hour for decorating. She works out how much to charge by using this formula:

Total charge in dollars = number of hours × 20

   a) How much does she charge for 10 hours of decorating?

   b) How much does she charge for 50 hours of decorating?

   c) Write a formula to calculate the cost, $C, for h hours of decorating.

5 Which is the correct formula for converting weeks (w) into days (d)?

   $w = 7d$ $d = 7w$ $d = 7 + w$ $d = w \div 7$

6 **a)** Write a formula for the number of seconds, s, in m minutes.

   b) Use your formula to find the number of seconds in 50 minutes.

7 **a)** Write a formula for the number of grams, x, in y kilograms.

   b) Use your formula to find the number of grams in 14 kilograms.

8 **a)** Write a formula to convert centimetres, c, to metres, m.

   b) Use your formula to convert 240 centimetres to metres.

9 There are 8 tomatoes in a can of tomatoes.

   a) Write a formula for the number of tomatoes, t, in c cans of tomatoes.

   b) Use your formula to find the number of tomatoes in 5 cans of tomatoes.

   c) If you need 24 tomatoes for a recipe, how many cans of tomatoes will you need?

10 Maya buys books from a website that charges $5 for postage and packing.

   a) Write a formula for her total bill, $t, if the cost of the books is $b.

   b) What is her total bill if she orders books that cost $35?

   c) Amir orders books from the same website. His total bill was $55. How much did his books cost?

11 Rahman is writing a formula to find the number of hours, h, in d days.

   He says, 'There are 24 hours in a day, so $24h = d$'.

   Rahman is not correct. Explain the mistake he has made and correct his formula.

12 The cost $c of buying b loaves of bread is $c = 2b$.

   a) What is the cost of each loaf of bread?

   b) The cost of some loaves of bread was $38. How many loaves of bread were bought?

13 The formula $c = 6k$ is used for calculating the cost, c, in dollars, of sending a parcel with mass k, kilograms.

a) Write a sentence explaining how to work out the cost of sending a parcel.

b) If a parcel cost \$42 to send, what was its mass?

14 Match the formulas to the situations:

$t = 5m$	Jane has m marbles, Tom has t marbles. Tom always has 4 more marbles than Jane.
$t = 3m$	The number of fence panels needed, t, is always one less that the number of fence posts needed, m.
$t = m - 1$	The cost, $\$t$, of buying m jars of jam that cost \$3 each.
$t = m + 4$	The total number of feet, t in a herd of m elephants.
$t = 4m$	The total number of pens, t, in m packs of pens when there are 5 pens in each pack.

Thinking and working mathematically activity

Start with the formula $t = 2m$. What situations could this formula describe?

Write down three different possible situations where this formula might apply.

Now make up a formula of your own, and two possible situations that it could describe.

Ask a partner what they think your formula is describing. Compare your answers.

Discuss

Why is it always important to state what your variables mean when you are using formulae?

Consolidation exercise

1 Copy and complete the input and output tables for these function machines.

a) $input \rightarrow \boxed{\text{add } 5} \rightarrow output$
b) $input \rightarrow \boxed{\times 4} \rightarrow output$
c) $input \rightarrow \boxed{\div 3} \rightarrow output$

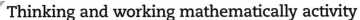

input	output
1	6
2	
6	
	18

input	output
1	4
2	
6	
	36

input	output
3	1
6	
12	
	10

2 For each of these functions, draw the function machine and the inverse function machine.

a) $x \mapsto x - 4$
b) $x \mapsto x \div 5$
c) $x \mapsto x + 7$
d) $x \mapsto 6x$

3 For the function $x \mapsto 7x$ what input would give these outputs?

a) 14
b) 35
c) 140
d) 0

4 Using the formula $m = d + 6$, find m when:

a) $d = 3$ b) $d = 14$ c) $d = 120$ d) $d = 53$

5 Which is the correct formula to find the number of millilitres, m, in x litres?

a) $m = \dfrac{1000}{x}$ b) $m = 1000 + x$ c) $m = 1000 - x$ d) $m = 1000x$

6 A packet contains 12 biscuits.

a) Write a formula for the total number of biscuits, b, in p packets.

b) Use your formula to find the number of biscuits in 4 packets.

7 Paul is laying a path. He knows that the width of the path will be 3 m. The length of the path will be L metres.

a) Write down a formula for the area, $a\,\text{m}^2$, of the path.

b) Use your formula to calculate the area of the path if it is going to be 12 metres long.

Paul is told that the cost of the path will be $8 per square metre.

c) Find the cost of his 12 m path.

d) What is the formula for the cost, c, of a path L metres long and 3 m wide?

Paul wants to put edging on each side of his path.

e) How much edging will he need for his 12 m path?

f) If the edging costs $5 per metre, how much will the edging to Paul's path cost?

g) Write a formula for the cost of the edging, e, for a path L metres long.

h) What is the total cost of Paul's path?

i) Can you find a formula for the total cost, t, of a path L metres long and 3 m wide?

> **Think about**
>
> If the width of the path in question 7 is changed, how would this change the formula?

End of chapter reflection

You should know that...	You should be able to...	Such as...
A function is a relationship where each input has a single output.	Generate outputs from a given function and identify inputs from a given output by considering inverse operations.	A function is $x \mapsto 4x$. Find: • the output when the input is 3 • the input when the output is 24.
A situation can be represented either in words or as a formula (single operation), and move between the two representations.	Construct and use a simple formula to find missing information.	Derive a formula for the number of hours, h, in d days and use it to calculate the number of hours in 8 days.

15 Area and volume

You will learn how to:
- Understand the relationships and convert between metric units of area, including hectares (ha), square metres (m²), square centimetres (cm²) and square millimetres (mm²).
- Derive and know the formula for the area of a triangle. Use the formula to calculate the area of triangles and compound shapes made from rectangles and triangles.
- Derive and use a formula for the volume of a cube or cuboid. Use the formula to calculate the volume of compound shapes made from cuboids, in cubic metres (m³), cubic centimetres (cm³) and cubic millimetres (mm³).
- Use knowledge of area, and properties of cubes and cuboids to calculate their surface area.

Starting point

Do you remember...

- how to use suitable metric measures?
  For example, which unit of measurement would be best to measure the height of a house?

- how to convert between different metric measures?
  For example, how many cm are in 16 m?

- how to multiply and divide by powers of 10?
  For example, divide 23 by 1000.

- how to find the area of a rectangle and the area of a right-angled triangle?
  For example, find the area of these shapes.

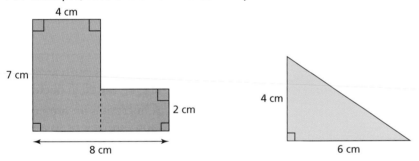

- how to find the surface area of simple cubes and cuboids from their nets?
  For example, find the surface area of this cuboid drawn on a centimetre grid.

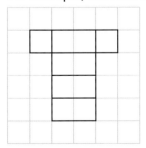

This will also be helpful when...

- you learn how to calculate the areas of other shapes and solve problems involving areas
- you learn how to find the volume of any prism and solve problems involving volume.

15.0 Getting started

How good are you at estimating area?

Decide in your class on a rectangular area to measure, like a table or a notice board. Ask your teacher to measure it for the class. This is your new measure of area, one table!

Each member of the class should estimate how many times the table or notice board will fit onto the floor. Now ask your teacher to measure the floor to find its area. Divide this by the area of the table or notice board to see how many would fit onto the floor. See who is the best estimator.

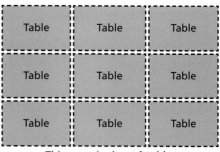

This room is about 9 tables

15.1 Converting between units of area

Thinking and working mathematically activity

- A square has sides measuring 4 cm. Find the area in cm^2.
- Convert 4 cm to millimetres and then find the area of the square in mm^2.
- Draw some rectangles and find their areas in cm^2 and mm^2.
- Make a conclusion about how to convert an area from cm^2 to mm^2.

Predict how to convert an area from m^2 to cm^2. Explain your prediction to a partner.

Conversion between units of area

$1\ cm^2 = 100\ mm^2$

$1\ m^2 = 10\ 000\ cm^2$

$1\ km^2 = 1\ 000\ 000\ m^2$

Key terms

Large areas, such as fields, are measured in **hectares** (ha).

A hectare is an area of land equivalent to a square measuring 100 m by 100 m.

1 hectare = 10 000 m^2.

Worked example 1

a) Convert 0.28 cm^2 to mm^2

b) Convert 13 600 cm^2 to m^2

a) $0.28\ cm^2 = 0.28 \times 100$
$= 28\ mm^2$

$1\ cm^2 = 10 \times 10$
$= 100\ mm^2$

To convert cm^2 to mm^2, you multiply by 100.

1 cm² = 100 mm²
1 cm = 10 mm
1 cm = 10 mm

b) $13\ 600\ cm^2$
$= 13\ 600 \div 10\ 000$
$= 1.36\ m^2$

$1\ m^2 = 100 \times 100$
$= 10\ 000\ cm^2$

To convert cm^2 to m^2, you divide by 10 000.

Area = 1 m² = 10 000 cm²
1 m = 100 cm
1 m = 100 cm

Did you know?

Hectares are used to measure areas of land.

Some countries also use acres as a unit of land area.

An acre is not a metric unit. One acre = 0.4 hectares, approximately.

Worked example 2

a) How many square metres are in 2.5 hectares?

b) A rectangular field is 300 m long and 200 m wide. How many hectares does the field cover?

a) $2.5 \times 10\,000 = 25\,000$ m^2	To convert hectares to metres you multiply by 10 000. $2.5 \times 10\,000 = 25\,000$	

2.5 ha

1 ha	1 ha	0.5 ha
10 000 m^2	10 000 m^2	5000 m^2

25 000 m^2

b) Area $= 300 \times 200$ $= 60\,000$ m^2 $60\,000 \div 10\,000 = 6$ ha	$A = $ length $\times$ width $300 \times 200 = 60\,000$ m^2 Now convert to hectares by dividing by 10 000.	

60 000 m^2

10 000 m^2	10 000 m^2	10 000 m^2	10 000 m^2	10 000 m^2	10 000 m^2
1 ha	1 ha	1 ha	1 ha	1 ha	1 ha

6 ha

Exercise 1

1 Copy and complete the following:

a) 3 m^2 = _____ cm^2 b) 25 m^2 = _____ cm^2 c) 0.5 m^2 = _____ cm^2

d) 200 cm^2 = _____ m^2 e) 4000 cm^2 = _____ m^2 f) 4500 cm^2 = _____ m^2

g) 200 mm^2 = _____ cm^2 h) 3000 mm^2 = _____ cm^2 i) 2.5 cm^2 = _____ mm^2

2 A piece of material is 1.8 m long and 1.5 m wide. Johan is cutting out squares of material, the sides of which are 10 cm. How many complete squares can he cut?

3 Maura's garden is a rectangle 6 m long and 5 m wide. The garden is grass except for the path which is a rectangle 600 cm long and 75 cm wide. What area of the garden is grass? Give your answer in m^2.

4 State whether the following are true or false. If they are false, explain the error.

a) 50 000 cm^2 = 5 m^2 b) 250 mm^2 = 2.5 cm^2

c) 2.5 m^2 = 25 000 000 cm^2 d) 0.45 ha = 45 000 m^2

5 Convert the following to square metres:

a) 2 ha   b) 10 ha   c) 3.5 ha   d) 0.25 ha

6 Convert the following to hectares:

a) 30 000 m^2 b) 1 000 000 m^2 c) 1000 m^2 d) 25 500 m^2

7 A sheet of metal is 1.2 m long and 1 m wide.

   a) What is the area of the sheet in cm^2?

   b) Each 10 cm^2 of the sheet has a mass of 40 g. What is the mass of the whole sheet in kg?

8 A rectangle has an area of 3500 cm^2. The rectangle is coloured in red and white. The area that is red is 0.097 m^2.

   Find the area in cm^2 that is white.

9 The floor of a room measures 3.2 m by 2.6 m.
   Babak wants to buy tiles that measure 20 cm by 20 cm to cover the floor.
   How many tiles are needed to cover the floor?

10 Jazmin has a rectangular room 8 m long and 5.5 m wide. The walls are 250 cm high. She wants to paint all four walls, but not the floor or the ceiling. A litre of paint will cover 14 m^2. Jazmin buys five litres of paint.

   Does she have enough paint to paint all four walls? Show how you found your answer.

15.2 Area of triangles

Thinking and working mathematically activity

- Copy this triangle onto centimetre square paper.

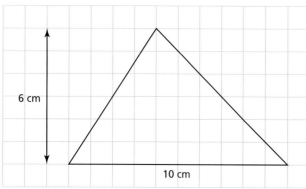

- Draw a rectangle around the triangle. Divide the triangle into two right-angled triangles, P and Q.
- Find the areas of triangles P and Q and the area of the original triangle. How does the area of the original triangle relate to the area of the surrounding rectangle?
- Investigate further with other triangles.

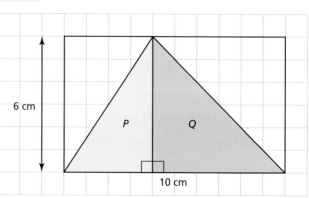

Worked example 3

Find the area of: **a)** 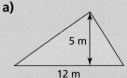 **b)**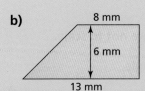

a) Area $= \frac{1}{2} \times$ base $\times$ height $= \frac{1}{2} (12 \times 5)$ $= \frac{1}{2} \times 60$ $= 30$ m²	The triangle is exactly half of the surrounding rectangle. The rectangle has base 12 m and height 5 m. Therefore, the area of the rectangle is 60 m² The area of the triangle is half the area of the rectangle. So, the area of the triangle is $= \frac{1}{2} \times 60$ $= 30$ m²	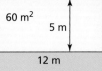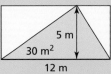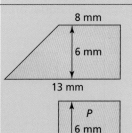
b) Area of rectangle $= 8 \times 6 = 48$ mm² Base of triangle $= 13 - 8 = 5$ mm Area of triangle $= \frac{1}{2} \times 5 \times 6 = 15$ mm² Total area $= 48 + 15 = 63$ mm²	The shape can be divided into a rectangle P and a triangle Q. Area of P is $8 \times 6 = 48$ mm² The base of Q is $13 - 8 = 5$ mm Area of Q is $\frac{1}{2} \times 5 \times 6 = 15$ mm² Total area $= 48 + 15 = 63$ mm²	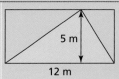

Area of a triangle

Area of a triangle $= \frac{1}{2} \times$ base $\times$ height

or Area $= \frac{1}{2} bh$

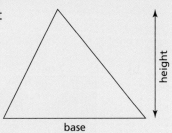

Exercise 2

The diagrams in this exercise are not to scale.

1 Find the area of each triangle.

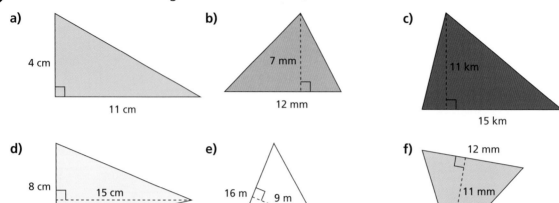

a) 4 cm, 11 cm

b) 7 mm, 12 mm

c) 11 km, 15 km

d) 8 cm, 15 cm

e) 16 m, 9 m

f) 12 mm, 11 mm

2 Match the shapes with equal areas. Give a reason for each match.

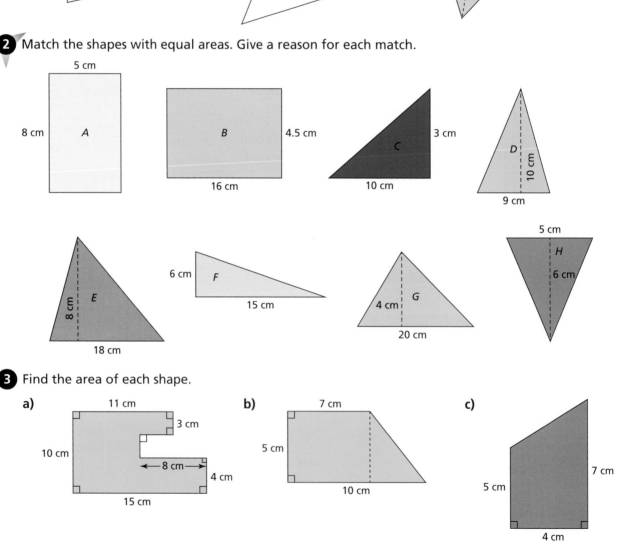

A: 5 cm, 8 cm

B: 16 cm, 4.5 cm

C: 10 cm, 3 cm

D: 9 cm, 10 cm

E: 8 cm, 18 cm

F: 6 cm, 15 cm

G: 4 cm, 20 cm

H: 5 cm, 6 cm

3 Find the area of each shape.

a) 11 cm, 3 cm, 10 cm, 8 cm, 4 cm, 15 cm

b) 7 cm, 5 cm, 10 cm

c) 7 cm, 5 cm, 4 cm

d)

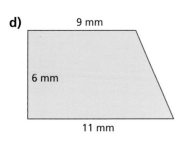

e)

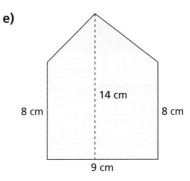

f)

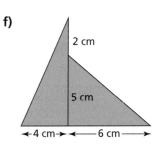

4 A triangle is drawn inside a rectangle. Find the shaded area.

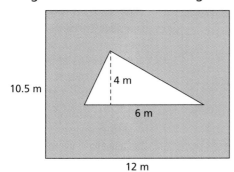

5 Copy and complete the table for different triangles.

Triangle	Area	Base length	Height
A		6 m	8 m
B	36 m²		3 m
C	108 mm²	36 mm	
D	24.5 cm²		7 cm

6 The triangle and the rectangle have the same area. Calculate the width, w, of the rectangle.

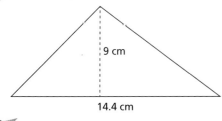

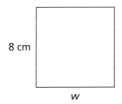

7 Show that one quarter of rectangle *ABCD* is coloured orange.

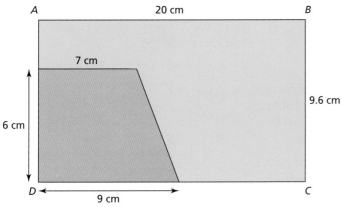

15.3 Volume of cuboids

 Thinking and working mathematically activity

You have 72 centimetre cubes.

- How many different cuboids can you build using all 72 cubes?
- Record the length, width and height of each of your cuboids in a table.

Length	Width	Height

- How can you combine the numbers in each row to make 72?
- Now try starting with 60 centimetre cubes. Find how many different cuboids you can make this time.

Key terms

The **volume** of a shape is how much space it occupies. This is often measured in cubic centimetres (centimetre cubes) which is written as cm^3. Other common measurements are cubic millimetres (mm^3) and cubic metres (m^3).

A volume does not have to be a whole number.

Worked example 4

a) How many 1-cm cubes make up the shape shown here?

b) What is the volume of a cuboid with width 2 m, length 7 m and height 3 m?

c) How can you work out the volume of a cuboid if you know its length, width and height?

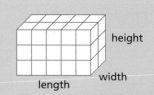

a) The base has $5 \times 2 = 10$ cubes. There are three layers so the total number of cubes is: $10 \times 3 = 30$ cubes	Start by finding the number of cubes in the base layer. There are three layers of cubes so you can multiply the number of cubes in one layer by 3 to find the total number of cubes.	In the base layer there are 2 rows of 5 cubes each. Number of cubes in the base layer $= 5 \times 2 = 10$ Three layers: $10 + 10 + 10 = 30$ cubes
b) The volume is: $2 \times 7 \times 3 = 42$ m^3	Find the number of cubes in the base layer. Multiply by the number of layers of cubes in the cuboid.	The number of metre cubes in the base layer is 2×7. You have 3 layers of 2×7 cubes. You can write this as: $2 \times 7 \times 3 = 42$ m^3 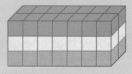
c) The volume is length × width × height	You can see from the examples above that the number of cubes to make up the base is found by multiplying length × width. To find the total number of cubes needed, multiply the number of cubes in the base by the height.	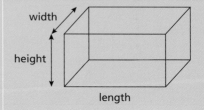

Exercise 3

Think about

Which of these cuboids is made from more cubes? How can you tell?

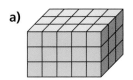

1 The following shapes are made up of centimetre cubes. Write down the length, width and height of each cuboid and work out its volume.

a) 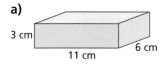   b) 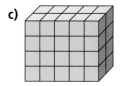   c) 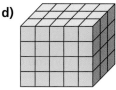   d)

2 Find the volume of these cubes in cm^3. The length of one side of the cube is given.

a) 1 cm   b) 2 cm   c) 5 cm   d) 10 cm

3 Calculate the volume of each cuboid.

a)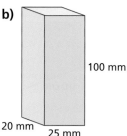
3 cm, 11 cm, 6 cm

b) 100 mm, 20 mm, 25 mm

c)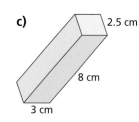
2.5 cm, 8 cm, 3 cm

d)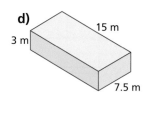
15 m, 3 m, 7.5 m

4 A cuboid has a square base of side 10 cm and is 8 cm high. What is its volume?

5 A cuboid is 5 mm long, 3 mm wide and 1 cm high. What is its volume in mm^3?

Tip

In question 5, make sure your length, width and height are in the same units before multiplying.

6 The first table gives the measurements of some cuboids. Match the cuboids to the volumes in the second table and find the missing width and the missing volume.

Cuboid	Length	Width	Height
A	3 cm	8 cm	4 cm
B	2 m	?	3 m
C	6 cm	6 cm	3 cm
D	4.5 cm	2 cm	5 cm
E	4 cm	3.5 cm	3 cm
F	2.5 mm	8 mm	2 mm
G	7 cm	3 cm	5 cm

Volume
45 cm^3
42 cm^3
96 cm^3
?
105 cm^3
96 m^3
108 cm^3

7 This cuboid has a volume of 280 cm^3.

Calculate its height.

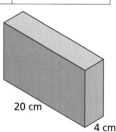
20 cm, 4 cm

8 Each of these cuboids has a volume of 360 cm³.

Calculate the length marked x in each one.

a)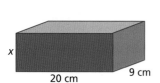

20 cm, 9 cm, x

b)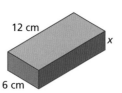

12 cm, 6 cm, x

c)

18 cm, x, 5 cm

d)

20 cm, 3 cm, x

9 Find the volume of each shape.

a)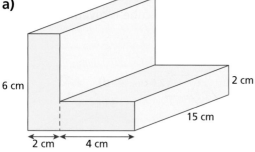

6 cm, 2 cm, 15 cm, 2 cm, 4 cm

b)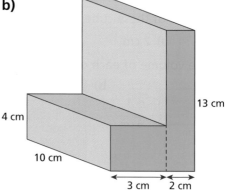

13 cm, 4 cm, 10 cm, 3 cm, 2 cm

c)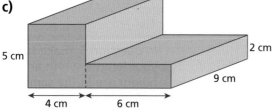

5 cm, 2 cm, 9 cm, 4 cm, 6 cm

d)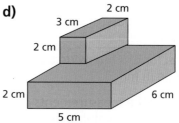

2 cm, 3 cm, 2 cm, 2 cm, 5 cm, 6 cm

e)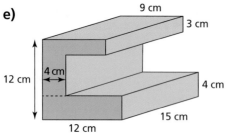

9 cm, 3 cm, 4 cm, 12 cm, 4 cm, 12 cm, 15 cm

f)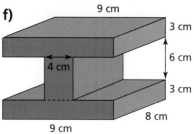

9 cm, 3 cm, 6 cm, 4 cm, 3 cm, 9 cm, 8 cm

> **Discuss**
>
> A six-sided dice has sides measuring 4 cm. Design a box that will hold exactly 60 of these dice.

170   Stage 7: Student's Book

10 Amy wants to find the volume of this cuboid.

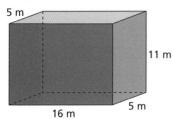

She says, 'All I need to do is multiply all the lengths together, so the volume is $16 \times 5 \times 11 \times 5 = 4400$ m^2.'

Do you agree with Amy? Explain your answer.

11 A cuboid has sides of 6 cm, 8 cm and 10 cm. It can be made using 480 cubes of side 1 cm. Samia has a set of cubes of side 2 cm. She says that she will need 240 of these to make the same cuboid.

Is she correct? Explain your answer.

15.4 Surface area of cuboids

Key terms

A **net** is a flat shape which can be folded to make a three-dimensional shape. The net of a cuboid could be folded to make the cuboid.

A solid can have more than one net. Here are two nets of the same cuboid, made up of the same six rectangles in different configurations.

The **surface area** is the area of the outside of a three-dimensional shape. In the case of a cuboid the surface area is the total area of all six faces.

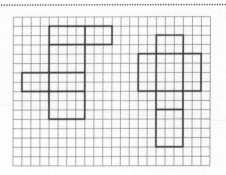

Did you know?

Not all combinations of the six rectangles make a net for the cuboid. This one is not a net for a cuboid as you could not fold it up to make a cuboid.

Can you see why?

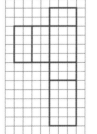

Worked example 5

Both of these cuboids a volume of 48 cm^3.

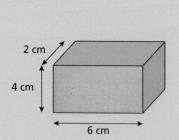 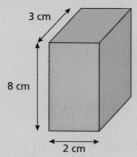

a) A net of the orange cuboid is drawn on squared paper. What is its surface area?

b) Calculate the surface area of the blue cuboid.

c) Do two cuboids with the same volume always have the same surface area?

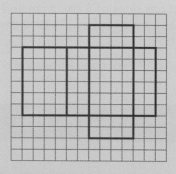

| a) Find the area of each face:
$2 \times 6 = 12$
$2 \times 6 = 12$
$4 \times 2 = 8$
$4 \times 2 = 8$
$4 \times 6 = 24$
$4 \times 6 = 24$

Surface area
$= 12 + 12 + 8 + 8 +$
$\quad 24 + 24$
$= 88$ cm^2 | Put the measurements on the net.

Check you have three pairs of faces.

Find the areas and add them up. | 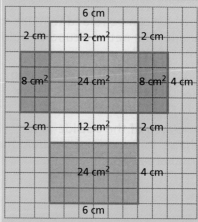 |

As the net is on 1 cm squared paper, you can add up the squares in each face to check.

There are 88 squares altogether, so the surface area is 88 cm^2

b) Find the area of each pair of faces. Front and back: $2 \times (2 \times 8) = 2 \times 16$ $= 32$ Right and left: $2 \times (8 \times 3) = 2 \times 24$ $= 48$ Top and bottom: $2 \times (2 \times 3) = 2 \times 6$ $= 12$ Surface area $= 32 + 48 + 12$ $= 92 \text{ cm}^2$	Look at each face on the cuboid. There are six faces: • Front and back • Right and left • Top and bottom Find the area of each face.	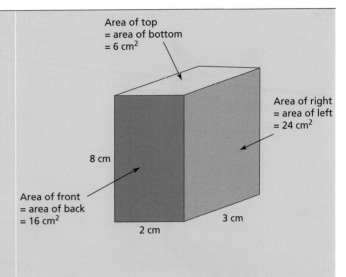
c) Two cuboids with the same volume do not necessarily have the same surface area. In fact, most of the time they won't.	In parts **a)** and **b)** you saw that the two cuboids of volume 48 cm³ had two different surface areas: 88 cm² and 92 cm².	

Exercise 4

1 These are nets of cuboids. Find the surface area of each one.

a)

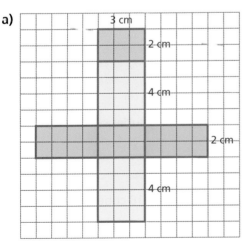

b)

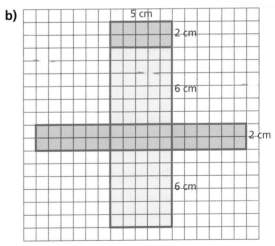

2 Calculate the surface area of each cuboid. You may want to sketch the net of the cuboid to help you.

a)

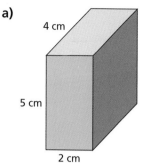

4 cm
5 cm
2 cm

b)

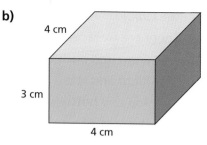

4 cm
3 cm
4 cm

c)

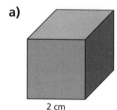

7 mm
2 mm
1 mm

d)

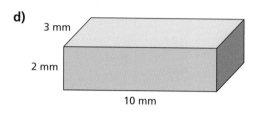

3 mm
2 mm
10 mm

3 Find the surface area of these cubes. You may want to draw a net to help you.

a)

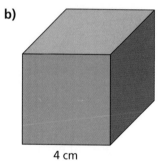

2 cm

b)

4 cm

4 A cube has side length s cm.

Sophia says that the surface area, A, of the cube is given by the formula $A = 6s^2$.

She is correct. Show how she got this expression.

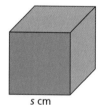

s cm

5 Use the formula $A = 6s^2$ to calculate the surface area of the following cubes.

a) Side length 11 cm   b) Side length 20 cm

6 Dara has 18 cubes each of side 1 cm.

He says, 'The greatest possible surface area of a cuboid made from all 18 cubes is when the cuboid has measurements 1 cm, 1 cm and 18 cm.'

Investigate to see if Dara is correct.

Thinking and working mathematically activity

- Cube *A* has side lengths of 4 cm. Calculate its surface area.
- The sides of cube *B* are twice as long as cube *A*. Find the surface area of cube *B*.
- How are the surface areas of the two cubes related?
- Explore what happens to the surface area of other cubes and cuboids when the side lengths are twice as long.

Consolidation exercise

1 Copy and complete the following.

a) $5 \text{ cm}^2 = $ _____ mm^2

b) $8 \text{ m}^2 = $ _____ cm^2

c) $12 \text{ km}^2 = $ _____ m^2

d) $400 \text{ mm}^2 = $ _____ cm^2

e) $30\,000 \text{ cm}^2 = $ _____ m^2

f) $700\,000 \text{ m}^2 = $ _____ ha

2 Place these areas in order of size, smallest first.

3500 m^2 7 ha $13\,000 \text{ m}^2$ 0.38 ha

3 A triangle has an area of 9 cm^2.

Sketch four possible triangles that it could be.

Label each triangle with its base and perpendicular height.

4 Calculate the area of the following shapes.

a)

b)

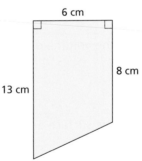

c)

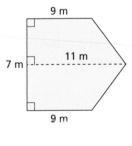

5 A room has a volume of 30 m^2. Its floor is a rectangle measuring 4 m by 3 m.
Find the height of the room.

6 A packet of biscuits measures 30 cm by 12 cm by 5 cm.

The packets of biscuits are packaged in large cardboard boxes measuring 1.5 m by 0.6 m by 0.6 m.

How many packets of biscuits fit inside each large cardboard box?

7 Is the following statement always, sometimes or never true? Explain your answer.

'The surface area of a cuboid is the area of one face multiplied by 6.'

8 A box used for packaging chocolates is a cube of side length 15 cm.

The box is printed with a design on all its faces that costs $0.05 per square centimetre.

How much does it cost to print each box?

9 Theo is told a cube has a surface area of 54 cm².

He says the volume must be 9 cm³.

Investigate Theo's statement.

End of chapter reflection

You should know that...	You should be able to...	Such as...
1 cm² = 100 mm² 1 m² = 10 000 cm² 1 km² = 1 000 000 m² 1 hectare = 10 000 m²	Convert between different units of area. Convert land area in m² into hectares.	**a)** Convert 7200 cm² to m². **b)** Convert 50 000 m² to hectares.
Area of a triangle $= \frac{1}{2} \times$ base $\times$ height height ◁ base	Calculate the area of a triangle. Calculate the area of shapes made from rectangles and triangles	Find the area of this shape: 5 cm, 8 cm, 9 cm
Volume of a cuboid = length × width × height	Calculate the volume of a cuboid.	Find the volume of a cuboid with length 3 cm, width 4 cm and height 5 cm.
The surface area of a cuboid is the sum of the area of all its faces. The surface area of a cube with side length s = $6s^2$	Calculate the surface area of a cuboid. Calculate the surface area of a cube.	**a)** Find the surface area: 15 cm, 5 cm, 8 cm **b)** Calculate the surface area of a cube with side length 8 cm.

16 Fractions, decimals and percentages

You will learn how to:
- Recognise that fractions, terminating decimals and percentages have equivalent values.
- Understand the relative size of quantities to compare and order decimals and fractions, using the symbols =, ≠, > and <.

Starting point

Do you remember...

- how to convert between simple fractions, percentages and decimals?

  For example, write 25% as a decimal and as a fraction in its simplest form.
- how to use the symbols < and >?

  For example, use < or > to write a correct statement about the numbers −4 and −7.
- how to compare and order decimals with one or two decimal places?

  For example, write these numbers in increasing order:
  0.27   0.3   0.19   0.1
- how to rewrite two fractions so that they have the same denominator?

  For example, write $\frac{2}{3}$ and $\frac{3}{4}$ with a common denominator.
- how to convert between improper fractions and mixed numbers?

  For example, write $\frac{15}{4}$ as a mixed number and write $2\frac{3}{5}$ as an improper fraction.

This will also be helpful when...

- you learn about recurring decimals and their equivalent fractions.

16.0 Getting started

Percentage means parts per hundred. For example, 77% means $\frac{77}{100}$.

Some whole-number percentages have equivalent fractions where the simplest form has denominator less than 100.

For example, $8\% = \frac{8}{100} = \frac{2}{25}$

- What is the denominator of the simplest form if the percentage is a prime number?
- What denominators are possible for fractions equivalent to whole-number percentages? Work systematically to find all of these. Try to explain your findings.
- Why is the denominator 3 not possible? Why is 75 not possible?

16.1 Equivalent fractions, decimals and percentages 1

Worked example 1

a) Convert 36% to a decimal and a fraction.
   Write the fraction in its simplest form.

b) Find the decimal and percentage equivalent of $\frac{3}{20}$.

a) $36 \div 100 = 0.36$ $\quad$ $36\% = 0.36$ $\quad$ $0.36 = \frac{36}{100}$ $\quad \div 4$ $\quad \frac{36}{100} = \frac{9}{25}$ $\quad \div 4$ $\quad 36\% = \frac{9}{25}$	To convert a percentage to a decimal, divide by 100. The smallest place value of 0.36 is hundredths. Write the fraction as 36 hundredths. Then simplify the fraction.
b) $\times 5$ $\quad \frac{3}{20} = \frac{15}{100}$ $\quad \times 5$ $\quad \frac{15}{100} = 0.15$ $\quad 0.15 \times 100 = 15$ $\quad 0.15 = 15\%$	A quick method is to rewrite the fraction with a denominator of 100. To convert a decimal to a percentage, multiply by 100.

Key terms

A **terminating decimal** is a decimal that stops after a number of decimal places, for example 0.6 or 0.74 or 0.31406.

Equivalent means equal in value. For example, 0.5, 50%, $\frac{5}{10}$ and $\frac{1}{2}$ are equivalent, because they all have the same value.

Exercise 1

1–9

1. Convert these percentages to decimals.
   a) 63% b) 70% c) 27% d) 9% e) 1%

2. Convert these terminating decimals to percentages.
   a) 0.65 b) 0.12 c) 0.9 d) 0.03 e) 0.4

3. Convert these terminating decimals to fractions. Write the fractions in their simplest form.
   a) 0.7 b) 0.8 c) 0.5 d) 0.44 e) 0.29 f) 0.22
   g) 0.35 h) 0.9 i) 0.09 j) 0.05 k) 0.01 l) 0.15

4. Jemima converts 0.13 to a fraction.

   She claims that 0.13 as a fraction is $\frac{1.3}{10}$

   Explain Jemima's mistake and correct her answer.

5 Convert these percentages to fractions. Write each fraction in its simplest form.

a) 30% b) 17% c) 24% d) 58% e) 95%

6 Convert each fraction to a decimal and a percentage.

a) $\frac{12}{100}$ b) $\frac{31}{100}$ c) $\frac{3}{100}$ d) $\frac{3}{10}$ e) $\frac{7}{10}$

f) $\frac{3}{4}$ g) $\frac{7}{25}$ h) $\frac{9}{20}$ i) $\frac{49}{50}$ j) $\frac{4}{5}$

7 Afia says that $\frac{12}{25}$ is equivalent to $\frac{24}{50}$.

Is Afia right? Explain your answer.

8 a) An advertisement says, '17 out of 20 people like *Butterfree*!' What percentage of people is this?

b) Yousef calculates that he spends 15% of his spare time playing computer games. What fraction of his spare time is this? Write the fraction in its simplest form.

9 Bella is converting $\frac{17}{25}$ to a decimal. Here is her working:

$\frac{17}{25} = \frac{17+75}{25+75} = \frac{92}{100} = 0.92$

She has made an error. Explain the error and correct the working.

10 **Technology question** Check how to use your calculator to convert between fractions, decimals and percentages.

Use your calculator to fill in the table. Write fractions in their simplest form.

Fraction	Percentage	Decimal
$\frac{3}{20}$		
	78%	
		0.05
$\frac{23}{25}$		

Thinking and working mathematically activity

For each calculation below, choose whether to give the answer as a fraction, decimal or percentage.

$80\% - \frac{3}{4}$ $0.32 + 21\%$ $0.5 + \frac{3}{20}$ $\frac{4}{5} - 25\%$ $\frac{4}{25} + 62\%$ $0.64 - \frac{11}{50}$

Discuss the methods you used. Is one method always the best, or are different methods better for different calculations?

Write your own questions like these, and show the most efficient way to answer each.

16.2 Equivalent fractions, decimals and percentages 2

Worked example 2

a) Find a fraction equivalent to 12.5%

b) Express $\frac{3}{8}$ as a decimal and as a percentage.

c) Write 1.46 as fraction and as a percentage.

a) $12.5\% = \frac{12.5}{100}$ $\frac{12.5}{100} = \frac{125}{1000}$ (× 10) $\frac{125}{1000} = \frac{25}{200} = \frac{1}{8}$ (÷ 5, ÷ 25) $12.5\% = \frac{1}{8}$	Write the percentage as the numerator and 100 as the denominator. Write an equivalent fraction that has the same value, with a whole number numerator. Simplify if possible.	**Alternative method** Convert the percentage to a decimal: $12.5 \div 100 = 0.125$ Then convert to a fraction: $\frac{125}{1000} = \frac{25}{200} = \frac{1}{8}$ (÷ 5, ÷ 25)
b) $\frac{3}{8} = 3 \div 8$ $8\overline{)3.^30^60^40}$ → 0.375 $\frac{3}{8} = 0.375$ $0.375 \times 100 = 37.5$ $0.375 = 37.5\%$	A fraction is a way of writing the result of a division. To convert to a decimal, divide the numerator by the denominator. To convert a decimal to a percentage, multiply by 100.	**Alternative method** Write an equivalent fraction that has numerator 10, 100 or 1000. $\frac{3 \times 125}{8 \times 125} = \frac{375}{1000}$ Write as a decimal: 0.375 (This method is not possible for some fractions.)
c) $0.46 = \frac{46}{100} = \frac{23}{50}$ So $1.46 = 1\frac{23}{50}$ $1.46 \times 100 = 146$ So $1.46 = 146\%$	First write 0.46 as a fraction in its simplest form. Then 1.46 can be expressed as a mixed number by adding in the whole number. To write 1.46 as a percentage, multiply it by 100.	

Exercise 2

1 Convert these terminating decimals to percentages.

    a) 0.009   **b)** 0.125   **c)** 0.075   **d)** 0.010   **e)** 1.75   **f)** 2.395

2 Convert these percentages to decimals.

    a) 23.8%   **b)** 6.7%   **c)** 20.5%   **d)** 139%   **e)** 190%   **f)** 2031%

3 Convert these terminating decimals to fractions. Write fractions larger than 1 as mixed numbers.

a) 0.999   b) 0.099   c) 0.008   d) 0.125   e) 0.025

f) 0.102   g) 0.005   h) 1.35   i) 4.2   j) 10.375

4 Convert these percentages to fractions. Write each fraction in its simplest form. Write fractions larger than 1 as mixed numbers.

a) 12.5%   b) 150%   c) 37.5%   d) 205%   e) 0.5%   f) 100.1%

5 a) Chen is converting 8.5% to a decimal. Here is his working:

$8.5\% = 0.85$

He has made an error. Explain the error and correct the working.

b) Lottie knows that $\frac{3}{4} = 75\%$

To find $\frac{3}{8}$, she multiplies 75% by 2.

Is Lottie correct? Explain your answer.

6 Write each fraction as a decimal and as a percentage.

a) $\frac{207}{100}$ b) $2\frac{1}{2}$ c) $\frac{3}{1000}$ d) $\frac{9}{200}$ e) $\frac{21}{250}$

f) $\frac{33}{1000}$ g) $\frac{13}{10}$ h) $\frac{21}{20}$ i) $\frac{333}{1000}$ j) $5\frac{1}{50}$

7 Use division to convert these fractions to decimals. Show your working.

a) $\frac{3}{5}$ b) $\frac{7}{8}$ c) $\frac{11}{40}$ d) $\frac{3}{80}$ e) $\frac{5}{16}$ f) $\frac{9}{16}$

8 Brindusa and Salma have collected some data about the favourite sports of students in their class.

Brindusa calculates that 35% like football best.

Salma calculates that $\frac{3}{8}$ of the students like hockey best.

Which sport is more popular in the class, football or hockey? Show your workings.

> **Think about**
>
> In question 8, if Brindusa and Salma's results are exact, what is the fewest possible number of students in the class?

9 Show that $1\frac{3}{25}$ is equivalent to 112%

10 a) Which of these fractions are between 0 and 0.25? Show how you know.

$\frac{3}{8}$ $\frac{3}{20}$ $\frac{9}{25}$ $\frac{7}{40}$ $\frac{3}{25}$

b) Can you think of any other methods to answer this question? Which method is the most efficient?

11 Seamus answers this question:

'What is 7 ÷ 8? Write your answer as a fraction.'

His answer is below.

$$\begin{array}{r} 0.875 \\ 8{\overline{)7.7^06^04^0}} \end{array}$$

$\frac{875}{1000} = \frac{175}{200} = \frac{7}{8}$

Is his answer correct?

Explain how to answer the question using less working.

> **Discuss**
>
> How many different methods can you think of to convert $\frac{8}{20}$ from a fraction to a percentage? Which is the most efficient method? Is this always the most efficient method for converting a fraction to a percentage?

Thinking and working mathematically activity

Convert these fractions to decimals. For each one, use the easiest method you can find.

$\frac{1}{8}$ $\quad$ $\frac{2}{8}$ $\quad$ $\frac{3}{8}$ $\quad$ $\frac{4}{8}$ $\quad$ $\frac{5}{8}$ $\quad$ $\frac{6}{8}$ $\quad$ $\frac{7}{8}$ $\quad$ $\frac{8}{8}$

Which fractions were easiest to convert? Explain why.

Show how to use the decimal equivalent of $\frac{1}{8}$ to find the decimal equivalent of each fraction below.

$\frac{3}{8}$ $\quad$ $\frac{7}{8}$ $\quad$ $\frac{1}{4}$ $\quad$ $\frac{1}{16}$

Thinking and working mathematically activity

Here are two pairs of fractions and their decimal equivalents:

$\frac{1}{16} = 0.0625$ $\qquad$ $\frac{11}{40} = 0.275$

$\frac{5}{8} = 0.625$ $\qquad$ $2\frac{3}{4} = 2.75$

Explain what you notice.

Find other pairs of fractions that fit the same pattern.

16.3 Comparing decimals and fractions

Key terms

Symbol	Meaning	Example
=	is equal to	$3 + 4 = 7$
≠	is not equal to	$3 + 4 \neq 6$
<	is less than	$6 < 7$
>	is greater than	$-6 > -7$

Worked example 3

a) Write < or > between the two numbers to make a correct statement.

$\quad$ 0.00645 $\qquad$ 0.0007

b) Write these numbers in order from smallest to largest.

$\quad$ $\frac{33}{20}$ $\qquad$ $\frac{19}{12}$ $\qquad$ $1\frac{19}{30}$

c) In the space below, which of the symbols =, ≠, < or > will make statement correct?

$\quad$ $\frac{3}{4}$ $\boxed{}$ $\frac{41}{50}$

a)		In the thousandths column, the 6 is bigger than the 0.	

a)

0	.	0	0	6	4	5
0	.	0	0	0	7	

In the thousandths column, the 6 is bigger than the 0.

$0.00645 > 0.0007$

0.001   0.005   0.0007   0.00645

b) $1\frac{13}{20}$ $1\frac{7}{12}$ $1\frac{19}{30}$

Rewrite as mixed numbers.

Alternative (longer) method

Convert to decimals using division:

1.65   1.58…   1.63…

$1\frac{39}{60}$ $1\frac{35}{60}$ $1\frac{38}{60}$

Write the fractions with a common denominator, and compare.

Write the decimals in order:

1.58…   1.63…   1.65

$\frac{19}{12}$ $1\frac{19}{30}$ $\frac{33}{20}$

Write in order from smallest to largest.

Write the fractions in order:

$\frac{19}{12}$ $1\frac{19}{30}$ $\frac{33}{20}$

c) $\frac{3}{4} = 0.75$

$\frac{41}{50} = \frac{82}{100} = 0.82$

Rewrite as decimals.

Alternative method

Write the fractions with a common denominator:

$\frac{3}{4} = \frac{75}{100}$ and $\frac{41}{50} = \frac{82}{100}$

$\frac{3}{4} \neq \frac{41}{50}$

$\frac{3}{4} < \frac{41}{50}$

Either $\neq$ or $<$ may be used to make the statement correct.

Exercise 3

1–7, 9–11

1 Which is the smallest number in each list?

a) 2.6   2.49   2.444

b) 1.2   1.122   1.109

c) 15.198   14.991   14.91

d) 1.43   1.399   1.322

2 Write each set of numbers in order, from smallest to largest.

a) 0.45, 0.4, 0.54, 0.47, 0.5

b) 0.601, 0.62, 0.6, 0.621, 0.612

c) 16.59, 16.6, 16.599, 16.95, 16.9

d) 0.037, 0.36, 0.04, 0.039, 0.4

3 In each statement there is a number missing. Write down a possible value for each.

a) $0.6 < \_\_\_ < 0.9$

b) $0.342 < \_\_\_ < 0.351$

c) $0.82 < \_\_\_ < 0.822$

d) $2.001 > \_\_\_ > 1.999$

e) $0.24 > \_\_\_ > 0.23$

f) $5.861 > \_\_\_ > 5.85$

4 In each space, write = or $\neq$ to make a correct statement.

a) $\frac{4}{5} \square \frac{12}{15}$

b) $\frac{2}{3} \square \frac{4}{5}$

c) $\frac{17}{8} \square 1\frac{7}{8}$

d) $1\frac{1}{4} \square \frac{16}{12}$

5 Use equivalent fractions to compare these pairs of fractions.

Write in the correct symbol <, > or =.

a) $\frac{5}{8} \square \frac{7}{16}$

b) $\frac{15}{14} \square \frac{2}{7}$

c) $\frac{3}{4} \square \frac{2}{3}$

d) $\frac{5}{6} \square \frac{7}{9}$

e) $\frac{5}{6} \square \frac{3}{4}$

f) $\frac{4}{9} \square \frac{5}{12}$

g) $\frac{9}{24} \square \frac{3}{8}$

h) $\frac{9}{14} \square \frac{13}{21}$

6 Convert to decimals to compare the fractions in each pair.

Write in the correct symbol, <, > or =.

a) $\frac{17}{20}$ ☐ $\frac{7}{8}$ b) $\frac{2}{5}$ ☐ $\frac{1}{8}$

c) $1\frac{1}{4}$ ☐ $\frac{29}{25}$ d) $1\frac{33}{40}$ ☐ $\frac{27}{16}$

7 Compare the fractions in each pair using <, > and =.

You can use equivalent fractions or convert to decimals.

a) $\frac{2}{3}$ ☐ $\frac{5}{12}$ b) $\frac{7}{8}$ ☐ $\frac{13}{16}$

c) $\frac{49}{25}$ ☐ $2\frac{1}{25}$ d) $1\frac{3}{4}$ ☐ $\frac{46}{25}$

8 For each pair of fractions, write which fraction has the greater value. Use a calculator to help you.

a) $\frac{3}{5}$ or $\frac{5}{8}$ b) $\frac{7}{16}$ or $\frac{11}{25}$

c) $\frac{7}{8}$ or $\frac{43}{50}$ d) $\frac{29}{32}$ or $\frac{91}{100}$

Think about

When is it easier to use equivalent fractions and when is it easier to convert to decimals?

Can you find some fractions that are difficult to order using equivalent fractions?

Can you find some fractions that are difficult to order when converting to decimals?

9 Write each set of fractions in order, from smallest to largest.

a) $\frac{2}{3}$ $\frac{7}{9}$ $\frac{5}{6}$ $\frac{13}{18}$ b) $\frac{16}{25}$ $\frac{29}{50}$ $\frac{47}{100}$ $\frac{7}{10}$

c) $\frac{17}{16}$ $\frac{5}{4}$ $1\frac{3}{16}$ $1\frac{1}{8}$ d) $\frac{19}{5}$ $3\frac{7}{10}$ $3\frac{13}{20}$ $\frac{7}{2}$

10 In each statement there is a number missing. Write down a possible mixed number for each.

a) $4\frac{1}{2}$ > ___ > 4   b) $1\frac{3}{4}$ < ___ < $2\frac{1}{8}$ c) $2\frac{1}{9}$ < ___ < $2\frac{1}{2}$

11 Find the missing number to make these statements true.

None of your answers should include an improper (top heavy) fraction.

a) $\frac{3}{4}$ < $\frac{☐}{8}$ b) $\frac{3}{4}$ < $\frac{☐}{6}$ c) $\frac{5}{8}$ > $\frac{☐}{3}$ d) $\frac{5}{7}$ < $\frac{☐}{8}$

Is there more than one correct answer for some of these questions?

Thinking and working mathematically activity

Amma is comparing the fractions $\frac{3}{8}$ and $\frac{1}{4}$. She says:

$\frac{3}{8}$ > $\frac{1}{4}$ because 3 > 1 and 8 > 4.

Is Amma correct?

Does her reasoning always work? Use examples to illustrate your answer.

Consolidation exercise  1–3, 6–9

1 a) There are 20 students in a class. Seven of the students wear glasses.
Write the percentage of students in the class who wear glasses.

b) Phillip scored $\frac{21}{40}$ in a test. Write his score as a percentage.

2 Match each percentage in the first row with the equivalent fraction or decimal in the second row.

Which percentage, and which fraction or decimal, are the odd ones out?

4%	140%	14%	4.1%	404%	4.4%	440%	144%	104%	10.4%
$4\frac{1}{25}$	$\frac{7}{50}$	10.4	$\frac{11}{250}$	$4\frac{2}{5}$	$\frac{26}{25}$	0.041	$\frac{1}{25}$	1.44	$\frac{7}{5}$

3 Below are the times taken by some athletes to run 100 metres.

Write the times in order, from the shortest to the longest.

10.689 s   10.709 s   10.899 s   10.688 s   10.75 s

4 Which of these two statements is true?

$\frac{3}{20} < \frac{5}{32}$ $\frac{3}{20} > \frac{5}{32}$

You can use a calculator to help you. Show your reasoning.

5 Maryann scored $\frac{19}{25}$ in a Music test, $\frac{11}{16}$ in a History test and $\frac{67}{80}$ in a Geography test.

  a) Express each mark as a decimal.

  b) In which subject did she get her best score?

6 **a)** Two bags contained the same number of oranges when they were full.
  One bag of oranges is now $\frac{5}{6}$ full and the other is $\frac{3}{4}$ full.

  Which bag contains more oranges?

  b) Jimmy's table tennis team won $\frac{13}{20}$ of their matches last year. This year they won $\frac{9}{15}$ of their matches.

  In which year was the team more successful? Explain your answer.

7 **a)** Write a fraction between $\frac{4}{7}$ and $\frac{5}{7}$. Show how you found your answer.

  b) Write a mixed number between $2\frac{3}{5}$ and $\frac{14}{5}$. Show how you found your answer.

End of chapter reflection

You should know that...	You should be able to...	Such as...
Fractions, decimals and percentages have equivalent values.	Convert between fractions, decimals and percentages	**a)** Write 1.85 as a fraction and a percentage. **b)** Write 22.5% as a fraction and as a decimal. **c)** Write $\frac{7}{8}$ as a decimal and as a percentage.
You can use the signs =, ≠, > and < to compare fractions and decimals.	Compare and order fractions and decimals.	Insert < or > to make correct statements: **a)** $\frac{5}{8}$ ☐ $\frac{16}{25}$ **b)** 0.4788 ☐ 0.48

17 Probability 1

You will learn how to:

- Use the language associated with probability and proportion to describe, compare, order and interpret the likelihood of outcomes.
- Understand and explain that probabilities range from 0 to 1, and can be represented as proper fractions, decimals and percentages.
- Identify all the possible mutually exclusive outcomes of a single event, and recognise when they are equally likely to happen.
- Understand how to find the theoretical probabilities of equally likely outcomes.

Do you remember...

- that a number can be written as a fraction, decimal or a percentage?

  For example, can you convert $\frac{3}{10}$ to a decimal and a percentage?
- how to place numbers on a number line?

  For example, can you place the numbers, 0.25, 0.75 and 0.5 on the line below?

  0 ———————————————————— 1
- the meaning of the words chance, likely, unlikely, certain and impossible?

  For example, can you write down two events that are certain to happen?

This will also be helpful when...

- you learn how to interpret and compare probabilities
- you learn how to find probabilities of combined events such as the probability of rolling a 6 or a 3 on a dice
- you learn more about listing outcomes of events to find probabilities.

17.0 Getting started

A game for two or more players

You will need:

- two dice
- a 3 by 3 grid like this for each player, filled in with numbers of your choice from 1 to 12 (no repeats)

12	5	9
2	11	7
8	4	10

Rules:

Players take turns to roll both dice and find the total of the scores.

Any player with this total on their grid may cross it out.

The winner is the first to cross out a horizontal, vertical or diagonal line.

- Play the game.
- Discuss your strategy with the other players before playing again. Consider how you can improve your performances.
- Are all the different scores equally likely to occur?
- If not, which are the most likely and which are the least likely? Are any of them certain? Are any of them impossible?
- Does it matter which number is in the middle of your grid?

17.1 The language of probability

Key terms

An **outcome** is a possible result of an **event**. For example, the outcomes of a football match are win, lose or draw. The event is the match. The outcomes from rolling a six-sided dice are the numbers 1, 2, 3, 4, 5 and 6. The event is rolling the dice.

A **probability** is the chance that a particular outcome, or result, will happen. When a fair coin is tossed, there are two possible outcomes because the coin has two different sides.

An event is described as **fair** when all the outcomes of the event are equally likely. When you spin a fair spinner, which is a regular hexagon with sides numbered from 1 to 6, then it is equally likely to land on any of the six numbers.

The **probability scale** from 0 to 1 represents probabilities visually. An outcome that is impossible has probability 0. An outcome with a probability of 1 is certain to happen. You can use a probability scale to compare probabilities.

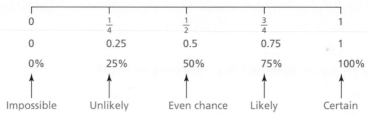

You can use fractions, decimals or percentages to express probabilities. For example, a coin toss has two possible and equally likely outcomes, so the probability of either outcome is one chance in two. You can write this as $\frac{1}{2}$ or 0.5 or 50%. These are different ways of writing the same thing.

Probabilities should not be written as ratios such as 1 : 4 or in words such as 1 in 5.

When items are selected **at random**, every item has the same chance of being selected.

> ### Did you know?
> Probabilities can never be greater than 1 or less than 0.

Worked example 1

Here is a probability scale:

impossible ———————————————————————— certain
0   1

a) On the scale, mark and estimate the probability that you will fall asleep before midnight tonight.

b) Describe something with a probability matching the arrow marked on the probability scale.

a) impossible ———————————— certain 0 1 The probability that you will fall asleep before midnight	It is very likely that you will fall asleep before midnight tonight – however, there is a small chance that you may not. You need to position your mark close to 1, but not at 1 because the event is not certain.
b) A bird will fly into your classroom tomorrow.	The arrow is pointing to a number close to 0. This means the probability is low. So, you are looking for an event that is very unlikely, but not impossible. For example, the arrow could represent the probability that a bird will fly into your classroom tomorrow. It's not impossible, but it is very unlikely.

Exercise 1

1 Choose one of the following probability words to describe the likelihood of each outcome happening.

Impossible   Unlikely   Even chance   Likely   Certain

a) getting tails when you toss a coin

b) you arriving at school on time on the next school day

c) a person chosen at random from the population of a large city being female

d) studying probability in maths next lesson

e) a day old baby talking in full sentences

f) the sun rising tomorrow morning.

2 a) Put the following outcomes in order from the least likely to the most likely.

– I will get homework tomorrow.
– It will snow tomorrow.
– I will eat breakfast tomorrow.

– It will rain tomorrow.
– It will be sunny tomorrow.

b) Use the terms below to describe the likelihood of each of the events in a) happening.

Certain   Very likely   Likely   Even chance   Unlikely   Very unlikely   Impossible

3 **a)** Describe an outcome which could be placed on the probability scale below at the places marked A, B, C and D.

b) Give an estimate of the probability of outcomes B and C as a decimal or fraction.

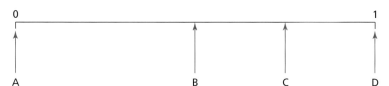

4 Here is a probability scale showing the probability that the 08:15 train will arrive late.

a) Write a sentence to describe how likely it is that the 08:15 train will be late.

b) The probability that an airplane from London arriving at 12:00 will be late is estimated as $\frac{3}{4}$.

Show this probability on a probability scale.

5 Sally has a pack of counters. She will select one without looking (at random).

Draw a probability scale and mark on it where each of the outcomes below should be.

a) Sally selects an orange counter. **b)** Sally selects a yellow counter.

c) Sally selects a green counter. **d)** Sally does not select an orange counter.

6 Bethan places the probabilities of five events happening on a scale.

Do you think she has placed the probabilities correctly?

> **Discuss**
>
> Discuss your answers to question 6 with a partner.

7 A dice is rolled. Decide if each statement is true or false. Give a reason for each answer.

a) It is very likely that the number will be greater than 4.

b) It is equally likely that the number will be prime as it will be an even number.

c) It is possible that the number will be a multiple of 6.

8 Sama has a bag that contains 11 red counters, 4 orange counters and 5 green counters. She takes out a counter at random and does not put it back in the bag. She then picks out another counter.

The second counter is twice as likely to be red as it is to be green. What was the colour of the first counter that was removed?

> **Tip**
>
> Remember that there is one less counter in the bag when the second counter is selected.

9 Tiago, Lucas and Will play a game where a card from the pack below is drawn at random.

1 2 3 4 5 6 7 8 9

Tiago wins if the number is odd.

Lucas wins if the number is prime.

Will wins if the number is a multiple of 3.

Is the game biased in favour of one player? Give a reason for your answer.

Discuss

Discuss with a partner how to make the game fair.

Thinking and working mathematically activity

Think of five different events that could fit into the boxes on this probability scale.

Discuss your answers with a partner.

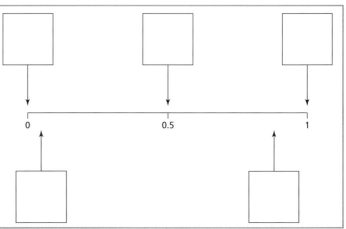

0   0.5   1

17.2 Listing outcomes

Key terms

Two outcomes are mutually exclusive if they cannot happen at the same time. For example, you cannot roll a 4 and a 6 on a single roll of a six-sided dice, so rolling a 4 and rolling a 6 are **mutually exclusive**.

Discuss

Can you think of different events that are mutually exclusive? Discuss with a partner.

Worked example 2

List all the possible mutually exclusive outcomes when the following events take place:

a) A coin is tossed

b) Spinning this spinner

a) heads or tails	There are two sides to a coin; heads and tails. There are no other possible outcomes from tossing a coin.	
b) blue, red or yellow	There are three colours, So, there are three possible outcomes.	

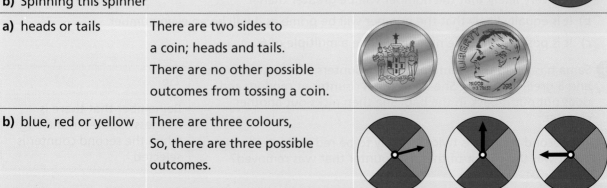

Think about

In part **b)** of Worked example 2, are the three outcomes of spinning the spinner equally likely?

1 List all the possible outcomes for rolling each of these dice.

a)

b)

six-sided dice   eight-sided dice

2 List all the possible outcomes for randomly selecting a ball from each bag.

a)   b)   c)   d)

3 List all the possible outcomes for choosing a letter at random from each of these words.

In each case, say whether the outcomes are equally likely or not.

a) M A T H S   b) E V E N T   c) S P I N N E R   d) P R O B A B I L I T Y

4 Are these pairs of outcomes mutually exclusive or not?

a) Going to school and not going to school.

b) Having toast for breakfast on a particular day and having cereal for breakfast on a particular day.

c) A student picked at random from a class being a hockey player and a student picked at random from a class being a football player.

5 Are these pairs of outcomes of a single roll of a six-sided dice mutually exclusive or not?

a) Rolling an even number

   Rolling an odd number

b) Rolling an even number

   Rolling a multiple of 3

c) Rolling an odd number

   Rolling a factor of 10

6 Sunita says the following:

'Rolling a prime number on a six-sided dice and rolling a square number on a six-sided dice are not mutually exclusive.'

Is Sunita correct? Explain your answer.

Tip

Write down all the prime numbers and the square numbers on a dice.

▼ Thinking and working mathematically activity

You have an eight-sided dice.

Give pairs of outcomes that are:

• mutually exclusive

• not mutually exclusive

Try to list as many as you can for each part.

Discuss your answers with other people in your class.

Key terms

When all the outcomes of an event are equally likely, you can calculate the probability of the event occurring as:

$$\frac{\text{number of favourable outcomes}}{\text{total number of outcomes}}$$

An event is described as **fair** if all the outcomes of the event are equally likely. When a fair spinner, which is a regular hexagon with sides numbered from 1 to 6, is spun then it is equally likely to land on any of the six numbers.

Using decimals and fractions to describe a probability

How likely is likely? Assigning a probability to a number means you can be more specific about how likely an outcome is to happen. Look at these two statements.

- I'm likely to win my next basketball match, and I'm likely to win my next tennis match.

- The probability I win my next basketball match is 0.7. The probability I win my next tennis match is 0.8.

Both statements say that I am likely to win both matches, but the second statement also tells me that I am more likely to win the tennis match.

> **Did you know?**
>
> You write **P(outcome)** to mean the probability of a particular outcome of an event.
>
> For example, P(it will rain tomorrow) means 'the probability that it will rain tomorrow'.

Worked example 3

Jess has a bag of discs.

She selects one at random.

Calculate the probability that:

a) Jess selects a red disc

b) Jess selects a green or an orange disc

c) Jess does not select a purple disc

d) Jess selects a blue disc.

a) There are 5 red discs out of a total of 12 discs so: $P(\text{red disc}) = \frac{5}{12}$	There are 5 red discs in the bag out of 12 discs altogether. Jess has 5 chances out of 12 to select a red or $\frac{5}{12}$	
b) There are a total of 4 green or orange discs out of a total of 12 discs so: $P(\text{green or orange disc})$ $= \frac{4}{12} = \frac{1}{3}$	There are 3 green discs and 1 orange disc, so altogether Jess has 4 chances out of 12 to select a green or orange disc or $\frac{4}{12}$ You could simplify this fraction to $\frac{1}{3}$	

c) 11 of the 12 discs are not purple so: P(not a purple disc) = $\frac{11}{12}$	There is 1 purple disc in the bag, so there are 11 that are not purple. Jess has 11 chances out of 12 to select a non-purple disc or $\frac{11}{12}$	
d) There are no blue discs in the bag of 12 discs so: P(blue disc) = 0	There are no blue discs in the bag so it is impossible for Jess to choose one.	

Exercise 3

1 Write down which of these events have equally likely outcomes.

- A counter is chosen without looking from a bag which has five red counters and five green counters.
- I spin this spinner.
- A letter is chosen from my name at random.

2 The eleven letters in the word PROBABILITY are written on eleven separate cards with one letter on each card. One card is chosen at random.

a) Write all the possible outcomes for this event.

b) Are there any outcomes that are more likely than others? Explain your answer.

> **Discuss**
>
> What is the probability of this spinner landing on yellow?
>
>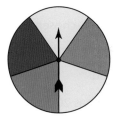
>
> What do you get when you add the probabilities of all the colours?

3 Esme rolls a fair six-sided dice.

Find the probability that she obtains:

a) a 5   **b)** a 1 or a 2   **c)** a number less than 6   **d)** an odd number

4 Jamal has some counters in front of him.

He says, 'The probability of selecting a red counter is $\frac{3}{5}$ because there are 3 reds and 5 blue counters'.

Do you agree with Jamal? Explain your answer.

5 A jar contains one gold coin, six silver coins and three copper coins and no other coins.

One coin is selected at random.

Find the probability that the coin selected is:

a) gold   **b)** silver   **c)** not gold   **d)** platinum

6 Felix spins the spinner shown.

Find the probability that the spinner stops on:

a) a blue sector

b) a sector with a negative number on it

c) a sector not coloured red

d) a yellow sector with an even number on it

e) a sector with a number greater than 3 on it

f) a sector with a number less than 20 on it.

7 A drawer contains 12 shirts. All the shirts are either plain, checked or striped. There are 7 plain shirts and 3 checked shirts.

A shirt is chosen at random.

Find the probability the shirt is:

a) checked   **b)** striped

8 There are 20 cards in a pack. All the cards are coloured either red, blue or green. 8 cards are coloured red. There are equal numbers of cards coloured blue and green.

A card is chosen at random.

Find the probability the card is:

a) red   **b)** green

9 Joran has 20 T-shirts. They are in different colours, some are plain and some have a pattern.

The table below shows how many of each type there are.

	Blue	Black	White
Plain T-shirt	3	5	4
Patterned T-shirt	3	2	3

Joran picks a T-shirt at random. Find the probability that the T-shirt is:

a) plain blue   **b)** patterned white   **c)** black

d) plain black or plain white   **e)** black or white

Write your answers as fractions.

10 Draw a spinner containing only five sectors, coloured red, yellow and green, which meets the following conditions:

$P(green) = \frac{3}{5}$

$P(yellow) = P(red)$

> **Tip**
>
> Make sure that all sectors are the same size.

11 Katy says, 'I am running for election as school captain and I have marked the probability of me winning on this probability scale. I can only win or lose, so I think I have an even chance of winning.'

equally likely to win or lose

Do you agree with Katy? Explain your answer.

▼ **Thinking and working mathematically activity**

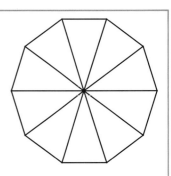

A game uses a spinner with 10 sections.
Each section can only be coloured yellow or red or green or blue.

1 Shade the spinner so that:

- the probability the spinner lands on a yellow section is 0

- the probability it lands on a red section is greater than the probability it lands on a green section

- the probability that the spinner lands on a blue section is more than 0.5.

2 How many possible ways are there to shade the spinner?
Convince a friend that you have found all the possible ways.

3 Design your own spinner and write down some sentences describing the likelihood of landing on different coloured sections.

Consolidation exercise

1 Choose words from the list below to help you describe the likelihood of each of these events.

impossible	likely	even chance	unlikely	certain

a) Fred has two marbles of the same size in a bag, a yellow one and a red one.
He puts his hand into the bag and, without looking, takes out a red marble.

b) Jaz has three marbles of the same size in a bag, a yellow, a blue and a red marble.
She puts her hand into he bag and, without looking, takes out a red marble.

c) Tom has six red marbles of the same size in a bag. He puts his hand into the bag and, without looking, takes out a green marble.

2 Which of these numbers could be probabilities?

$\frac{2}{3}$ $\frac{5}{4}$ -0.5   12%   1.4

3 A disc is selected at random from a set of discs. Which of the following sets of discs gives the highest probability of selecting a red disc?

a)   b)    c)   d) ●●●●●●

4 Which is the odd one out?

A: the probability of a multiple of 3 showing when a fair dice is rolled

B: the probability of spinning a 3 on the spinner shown

C: The probability a month, selected at random, ending in the letter 'y'

D: the probability of a football match ending in a draw

E: the probability of selecting a blue counter at random from the counters shown

5 Three players take part in a game. One card is chosen at random from these nine cards.

| 1 | 2 | 2 | 2 | 3 | 4 | 5 | 5 | 6 |

Player 1 wins if the number chosen is 2.

Player 2 wins if the number chosen is a square number.

Otherwise Player 3 wins.

a) Explain why the game is unfair.

b) Change the number on one of the cards to make the game fair.

6 Padma has 20 cards.

One card is picked at random.

Find examples:

a) of two outcomes that are equally likely

b) of an outcome that is almost certain

c) of an outcome that is impossible.

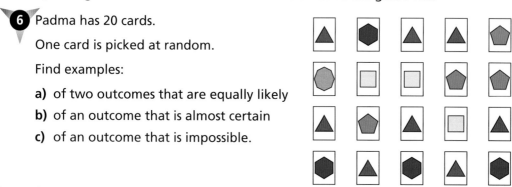

7 Which of these pairs of events are mutually exclusive? Give a reason for your answers.

a) winning a football match and drawing the same match

b) wearing one red sock and one blue sock

c) eating toast for breakfast and chips for dinner

d) being on time and being late one day at school.

8 Nadia has a bag of marbles. She says the probability of randomly picking a blue marble is $\frac{3}{8}$. Gareth says there must be eight marbles in the bag.

Explain why Gareth mightnot be correct.

End of chapter reflection

You should know that...	You should be able to...	Such as...
The probability of a particular outcome of an event can be described using probability words such as certain, likely, even chance, unlikely and impossible.	Describe the probability of an outcome in words.	Describe the probability that it will rain tomorrow.
You can tell how likely an outcome is from its position on the probability scale.	Describe how likely an outcome is from its position on the probability scale.	The arrow shows the probability of tomorrow being sunny. How likely is it to be sunny tomorrow? 0 ——————↑—— 1

An outcome is the possible result when an event takes place.	Identify all the possible mutually exclusive outcomes of a single event.	List all the possible outcomes of rolling a ten-sided dice.
When the outcomes of an event are equally likely, you can calculate the probability as: $\dfrac{\text{number of favourable outcomes}}{\text{total number of outcomes}}$	Calculate the probability of an event with equally likely outcomes.	Tracey has a jar of buttons. There are nine black buttons, five brown buttons and one white button. Tracey selects a button at random. Calculate the probability that Tracey selects: **a)** a brown button **b)** a button that is not black.

Transformations

You will learn how to:

- Use knowledge of 2D shapes and coordinates to find the distance between two coordinates that have the same x or y coordinate (without the aid of a grid).
- Use knowledge of translation of 2D shapes to identify the corresponding points between the original and the translated image, without the use of a grid.
- Reflect 2D shapes on coordinate grids, in a given mirror line (x- or y-axis), recognising that the image is congruent to the object after a reflection.
- Rotate shapes 90° and 180° around a centre of rotation, recognising that the image is congruent to the object after a rotation.
- Understand that the image is mathematically similar to the object after enlargement. Use positive integer scale factors to perform and identify enlargements.

Starting point

Do you remember...

- what coordinates are and how to plot them on a coordinate grid?

  For example, the points plotted on the grid have coordinates
  $A(2, 1)$, $B(5, 3)$, $C(-1, 2)$, $D(-3, 1)$, $E(-3, -2)$, $F(4, -3)$

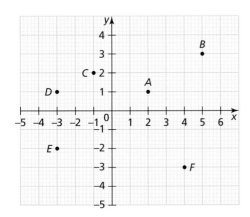

- translating a shape on a coordinate grid?

  For example, translating shape C by 2 units to the left and 5 units up puts it in the position shown by shape D.

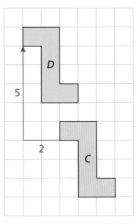

- reflecting a shape in a line?

  For example, shape A has been reflected in the mirror line to shape B.

  Notice the reflection is perpendicular to the mirror line.

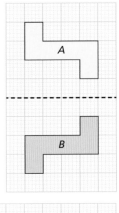

- rotating a shape around a point?

  For example, shape B has been rotated 90° clockwise around a point on B, to shape C.

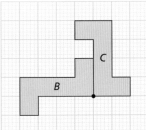

- you transform shapes using a combination of transformations
- you describe a transformation that maps one shape to another.

18.0 Getting started

Rangoli patterns are an art form, originating from India.

Rangoli patterns usually have either lines of symmetry or rotational symmetry. Here are some examples of Rangoli patterns:

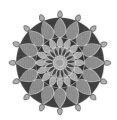

> **Did you know?**
>
> Rangoli patterns are often created on the floor in living rooms or courtyards to welcome people.
>
> They are made using materials such as coloured rice, dry flour, coloured sand or flower petals.

Create your own Rangoli pattern by following the steps below.

- Draw a square grid measuring 10 squares by 10 squares. Divide the grid in half horizontally and vertically.
- Draw a pattern with straight lines in the top left hand corner.
- Then reflect your pattern in both the horizontal and vertical lines to create a pattern with two lines of symmetry.
- Colour this pattern to create your own Rangoli pattern. Your colours should also be symmetrical.

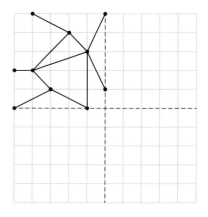

18.1 Translations

Key terms

A **translation** is a movement of a shape. You usually state the distance and direction of the movement horizontally (left/right) and vertically (up/down).

With a translation, the original shape is called the **object**, and the translated shape is called the **image**.

 Thinking and working mathematically activity

On this grid, A, B, C and D are connected to make a rectangle.

- What are the coordinates of points A, B, C and D?
- What are the lengths of sides AB, BC, CD and AD?
- Could you work out these lengths from the coordinates, without the diagram?
- A different rectangle $EFGH$ also has sides that are parallel to the axes. The coordinates of E are $(7, 3)$. EF has length 4 units. Find the coordinates of the possible positions of F.

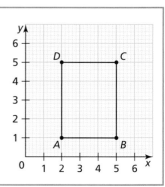

Worked example 1

The shape *ABCDEF* is shown (not drawn to scale).

All its sides are parallel to the axes.

The coordinates of *A* are (3, 4)

The coordinates of *E* are (5, 6)

The coordinates of *C* are (7, 10)

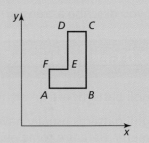

a) What are the coordinates of *D* and *B*?

b) Find the length of side *BC*.

c) The shape is translated 6 units right and 7 units up.
   Write down the new coordinates of *A*.

a) *D* is (5, 10) *B* is (7, 4)	*D* has the same *x*-coordinate as *E*. This is 5. *D* has the same *y*-coordinate as *C*. This is 10. *B* has the same *x*-coordinate as *C*. This is 7. *B* has the same *y*-coordinate as *A*. This is 4.	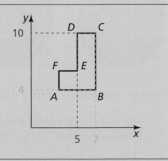
b) *BC* is 6 units	*C* is given as (7, 10) *B* is (7, 4), found in part **a)**. *B* and *C* have the same *x*-coordinate. The difference between the *y*-coordinates is $10 - 4 = 6$ units.	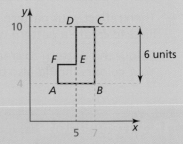
c) Image of *A* is (9, 11)	*A* is (3, 4) Translate 6 units right and 7 units up to give the image of *A* as: $(3 + 6, 4 + 7) = (9, 11)$	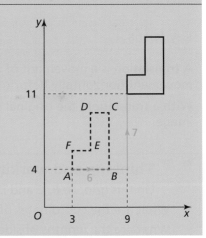

1 *ABC* is a right-angled triangle.

The line *AB* is parallel to the *x*-axis.

a) The distance between *A* and *B* is 6 units.
Write down the coordinates of *B*.

b) The distance between *B* and *C* is 14 units.
Write down the coordinates of *C*.

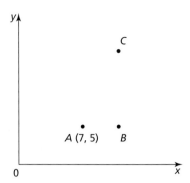

2 Write down the distances between points with the following coordinates.

a) (3, 7) and (8, 7)　　**b)** (2, 5) and (2, –4)　　**c)** (–1, 6) and (3, 6)

d) (4, 3) and (4, 9)　　**e)** (7, 3) and (7, 11)　　**f)** (5, 8) and (12, 8)

3 The distance between points *A* and *B* is 7 units.

The coordinates of *A* are (2, 9).

Write down possible coordinates for point *B*.

4 Write each pair of points in the correct position in the table.

a) (3, 5) and (3, 9)　　**b)** (–2, –4) and (–2, –1)　　**c)** (1, 5) and (0, 5)

d) (0, –1) and (7, –1)　　**e)** (0, –2) and (0, 7)　　**f)** (–5, 3) and (–9, 3)

	On a vertical line	On a horizontal line
Length between points < 4 units		
Length between points = 4 units		
Length between points > 4 units		

5 Write down the image of the point (5, –3) after a translation of:

a) 6 units down　　**b)** 5 units to the left　　**c)** 3 units up　　**d)** 8 units to the right.

6 Write down the image of the point (–2, 4) after a translation of:

a) 2 units right and 3 units down　　**b)** 3 units left and 5 units up

c) 5 units left and 7 units down　　**d)** 6 units right and 2 units up.

7 The coordinates of the vertices of triangle *ABC* are *A*(0, 3), *B*(–2, 4) and *C*(1, 1).

The triangle is translated 3 units right and 4 units down.

Write down the vertices of the image.

8 A translation maps the point (–1, 1) to the point (3, 6).

Copy and complete this description of the translation:

Under this transformation, the point moves _____ units right and 5 units _____ .

9 The image of the point (3, –5) under a translation is (2, 2).

Find the coordinates of the image of point (0, 4) under the same translation.

10 Match each pair of points to the correct translation.

(3, 6) moves to (4, 9)	2 right and 1 up
(–1, 4) moves to (1, 5)	1 right and 3 up
(4, –3) moves to (3, 6)	4 left and 3 down
(3, –2) moves to (–1, –5)	1 left and 9 up

11 *ABCD* is a parallelogram. The vertices of three of its coordinates are *A*(2, 3), *B*(7, 3) and *C*(8, 5).

a) Write down the coordinates of point *D*.

b) Write down the translation that moves point *A* to point *D*.

c) What is the length of *AB*?

12 *ABC* is an isosceles triangle. *A* has coordinates (3, 4). *B* has coordinates (5, 8).

Write down two possible coordinates for point *C*.

Give reasons for your answer.

13 Theo drew a parallelogram *ABCD* with *A*(2, 3) and *C*(7, 9).

Leo said, 'If you tell me the coordinates of *D*, then I can tell you the coordinates of *B*.'

Is Leo correct? Give reasons for your answer.

18.2 Reflections and rotations

Key terms

A **reflection** is a mirroring or flipping of a shape in a line. A reflection always has a **mirror line**.

A **rotation** is the turning of a shape. You usually state the angle and direction of the rotation as well as the point that the shape turns around. This point is called the **centre of rotation.**

Translations, reflections and rotations are called **transformations** because they transform or change one shape into another.

A rotation takes place when a shape has turned around a centre of rotation.

The effects can be seen using tracing paper:

- Trace the triangle *ABC* onto a piece of tracing paper.

- Put the sharp end of your pencil on the 'centre of rotation' and turn the tracing paper until the triangle falls exactly on top of the image.

- How many degrees have you turned the paper through? It should be either 90° or 180°.

- Which of the 90° rotations is clockwise, which is anticlockwise?

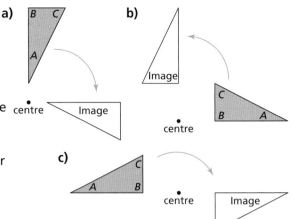

Worked example 2

a) Reflect shape *A* in the *x*-axis and label the image *B*.

b) Reflect shape *A* in the *y*-axis and label the image *C*.

c) Which of the images are congruent to *A*?

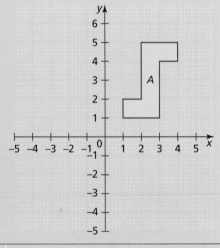

a)

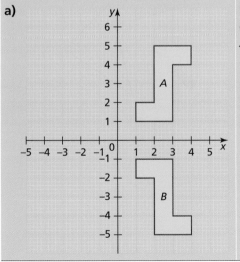

Reflect each of the corners of the shape *A* in the *x*-axis and join them up.

Use lines that are perpendicular to the mirror line when reflecting.

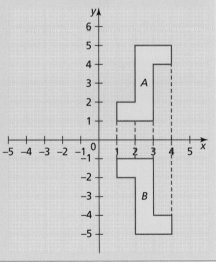

b)

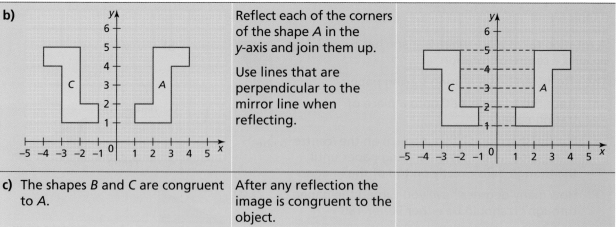

Reflect each of the corners of the shape A in the y-axis and join them up.

Use lines that are perpendicular to the mirror line when reflecting.

c) The shapes B and C are congruent to A.

After any reflection the image is congruent to the object.

Worked example 3

Rotate triangle T by 90° about point O by 90° in a clockwise direction.

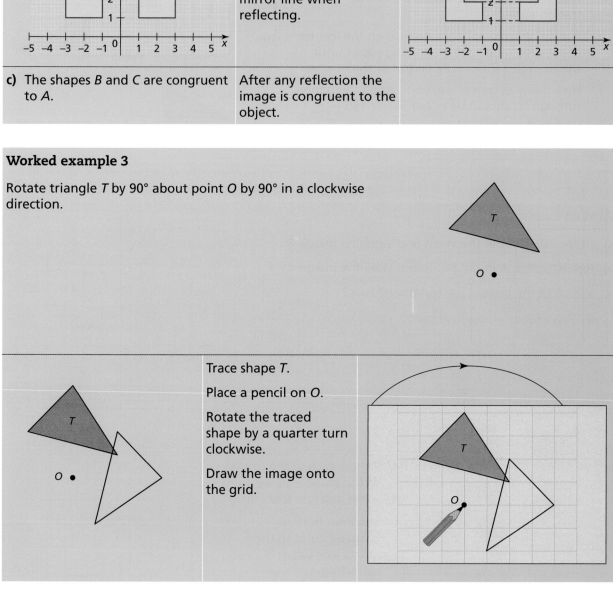

Trace shape T.

Place a pencil on O.

Rotate the traced shape by a quarter turn clockwise.

Draw the image onto the grid.

1 Copy shape *A* onto squared paper.

a) Reflect shape *A* in the *y*-axis. Label the image *B*.

b) Reflect shape *A* in the *x*-axis. Label the image *C*.

c) Reflect shape *B* in the *x*-axis. Label the image *D*.

d) Which of the shapes are congruent to *A*?

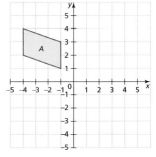

2 Copy shape *A* onto squared paper.

a) Reflect shape *A* in the *x*-axis. Label the image *B*.

b) Reflect shape *A* in the *y*-axis. Label the image *C*.

c) Reflect shape *B* in the *y*-axis. Label the image *D*.

d) Which shape is *D* a reflection of in the *x*-axis?

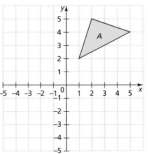

3 Three brothers looked at this diagram.

a) Bri said shape *B* is a reflection of shape *A* in the *x*-axis. Is he correct? Explain your answer.

b) Mal said shape *D* is a reflection of shape *A* in the *y*-axis. Explain how you know that Mal is not correct.

c) Kev said shape *C* is a reflection of shape *B* in the *y*-axis. Explain how you know that Kev is not correct.

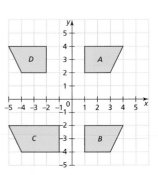

4 Draw each shape below onto squared paper then rotate it through 90° clockwise about the centre of rotation and draw its image. Use tracing paper if it helps you.

a)

b)

c)

5 Draw each shape below onto squared paper then rotate it through 90° anticlockwise about the centre of rotation and draw its image. Use tracing paper if it helps you.

a)

b)

c)

6 Draw each shape below onto squared paper then rotate it through 180° about the centre of rotation and draw its image. Use tracing paper if it helps you.

a)

b)

c)

7 Rotate the shape anticlockwise about the centre of rotation O by the angle shown.

a) 90°

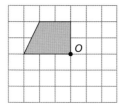

b) 180°

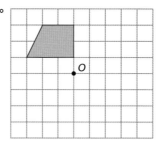

c) 270°

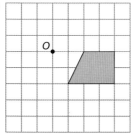

d) 90°

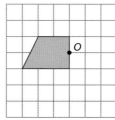

8 Here is a letter F shape.

Write down which of these shapes could *not* be a rotation of the F shape.
Give a reason for each answer.

Shape P

Shape Q

Shape R

9 Copy shape Q onto squared paper and reflect it in the x-axis.

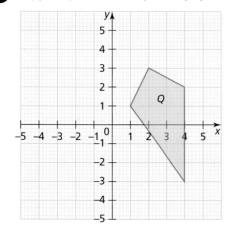

10 Copy shape R onto squared paper and reflect it in the y-axis.

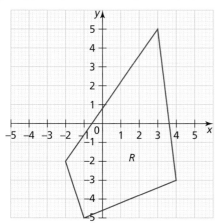

11 Draw each shape onto squared paper.

Rotate each shape about the marked centre using the given angle of rotation.

a)

90° clockwise

b)

90° anticlockwise

c)

180°

▼ Thinking and working mathematically activity

Look at the triangle ABC and its image $A'B'C'$.

Triangle ABC has been reflected to give triangle $A'B'C'$.

- What is the mirror line of this reflection?

- Investigate the connection between the coordinates of A, B and C and the coordinates of their images.

- Will the same connection hold if any point is reflected in this same mirror line? Investigate.

- Now consider what happens when you reflect a shape in the y-axis. Investigate the relationship between the coordinates of an object and an image in this case.

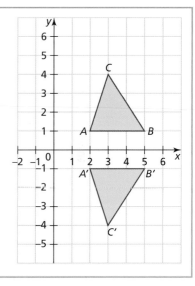

18.3 Enlargements

Key terms

You make an **enlargement** when you transform a shape to make it either bigger or smaller.

The image will be **similar** to the original shape. This means that the object and the image will have the same shape.

The **scale factor** determines the size of the **image**.

Worked example 4

Enlarge each shape with the given scale factor.

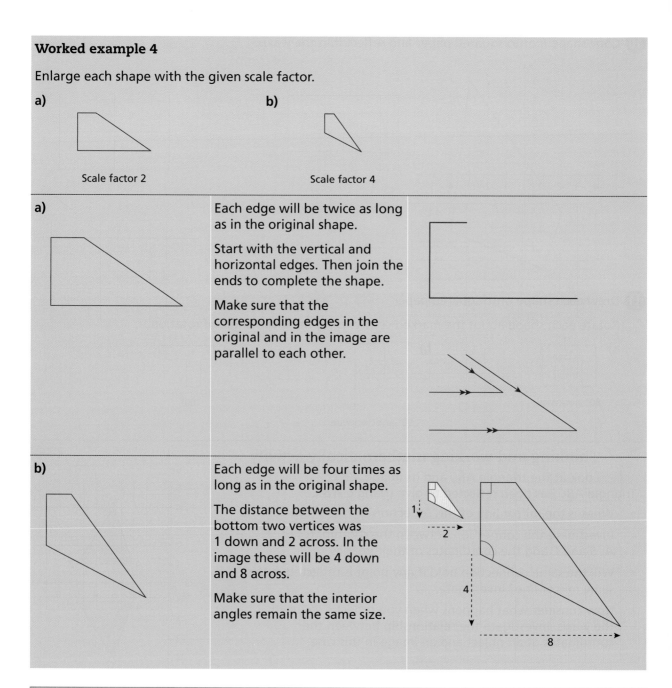

a)

Scale factor 2

b)

Scale factor 4

a)	Each edge will be twice as long as in the original shape. Start with the vertical and horizontal edges. Then join the ends to complete the shape. Make sure that the corresponding edges in the original and in the image are parallel to each other.	
b)	Each edge will be four times as long as in the original shape. The distance between the bottom two vertices was 1 down and 2 across. In the image these will be 4 down and 8 across. Make sure that the interior angles remain the same size.	

▼ Thinking and working mathematically activity

- Draw a triangle on squared paper. Enlarge the shape using a scale factor of 2.
- Measure the angles in the object and the image. What do you notice?
- Draw other triangles and enlarge them. Investigate the sizes of the angles in each object and its image.
- Form some conclusions about what happens to the angles in a shape when it is enlarged.

1 a) Write the letters of the shapes that are **not** enlargements of the original rectangle A.

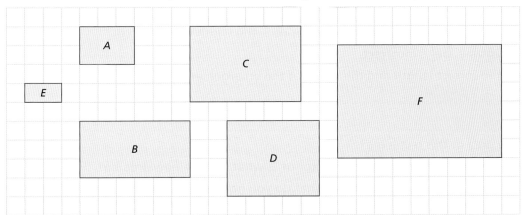

b) Identify the scale factor of any of the shapes which are an enlargement of A.

> Tip
>
> When enlarging a slanted line, count the number of squares it covers horizontally and vertically and multiply this by the scale factor for the new line.
>
> For example, 2 right and 1 down becomes 6 right and 3 down when using a scale factor of 3.
>
> Notice that the corresponding lines are parallel.

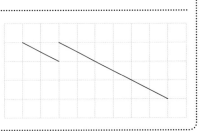

2 The following shapes are drawn onto cm squared paper.

Enlarge each shape by the scale factor given.

a)

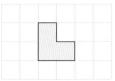

Scale factor 2

b)

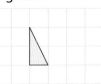

Scale factor 3

c)

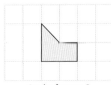

Scale factor 3

d)

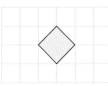

Scale factor 4

e)

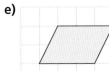

Scale factor 2

f)

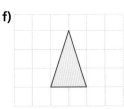

Scale factor 3

> Discuss
>
> What would the image look like if you enlarged a rectangle by a scale factor of 1?

3 The quadrilateral is drawn onto cm squared paper.

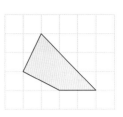

  a) Enlarge the shape by a scale factor of 2.

  b) Enlarge the shape by a scale factor of 3.

4 Eden draws a rectangle with length 3 cm and width 2 cm.

  Joel says, 'If you enlarge it by a scale factor of 5, the width will be 15 cm.'

  a) Explain why Joel is incorrect.

  b) Find the correct length and the width of the enlargement.

Think about

What is the area of this rectangle?

If you enlarge the rectangle by a scale factor of 2, what will the new area be?

Try this out with different scale factors. What do you notice?

2 cm 3 cm

---

5 Write down if each statement is always, sometimes or never true.

  a) An enlargement makes a shape bigger.

  b) When a shape is enlarged the angles remain the same size.

  c) An enlargement of scale factor 2 makes each edge 3 times longer.

6 Trinity enlarged a shape by a scale factor of 4.

  She drew the following image onto cm squared paper.

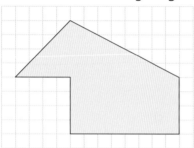

  Draw the original shape before the enlargement.

7 The diagram shows shape L and two enlargements.

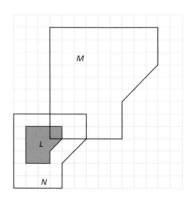

  a) What is the scale factor from shape L to shape M?

  b) What is the scale factor from shape L to shape N?

  c) Explain why shapes L and M are similar.

  d) Ollie says the scale factor from shape N to shape M is 2.
  Explain why Ollie is not correct.

Consolidation exercise

1 Harry says that the distance between (3, 9) and (3, 21) is the same as the distance between (–9, –8) and (–9, 2). Show that Harry is not correct.

2 The coordinates of the vertices of triangle ABC are A(–2, 5), B(3, 6) and C(4, –1). The triangle is translated 3 units right and 4 units down. Write down the coordinates of the vertices of the image.

3 The image of the point (5, –3) under a translation is (1, 3). Find the coordinates of the image of (–1, 4) under the same translation.

4 Copy the shapes below onto squared paper.

a) Reflect shape P in the y-axis. Label the image Q.

b) Reflect shape ABCD in the y-axis. Label the image A'B'C'D'.

c) Which of the shapes are congruent to shape P?

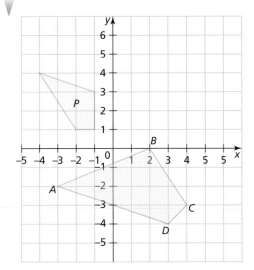

5 Draw each shape below onto squared paper. Rotate each shape through the angle shown about the centre of rotation.

a)

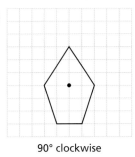

90° clockwise

b)

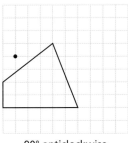

90° anticlockwise

c)

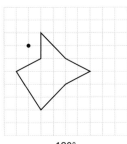

180°

6 Copy each shape onto squared paper.

Enlarge each shape by the scale factor given.

a)

Scale factor 3

b)

Scale factor 4

7 Write down if each statement is always, sometimes or never true.

a) When a shape is rotated, the image is congruent to it.

b) When a shape is enlarged, the image is congruent to it.

8 Write down the scale factor for the following enlargements.

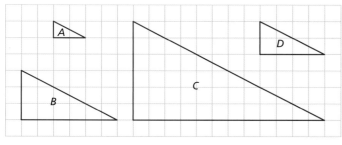

a) *A* to *B*　　　　b) *A* to *C*　　　　c) *A* to *D*　　　　d) *D* to *C*

e) Ollie draws two arrows side by side as shown.

He says, 'I have enlarged the left arrow by a scale factor of 2'.

Ben says that Ollie has made a mistake.

Who is correct? Explain your answer.

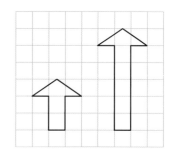

End of chapter reflection

You should know that...	You should be able to...	Such as...
The image is congruent to the object after a translation.	Find the distance between two coordinates that have the same *x*- or *y*-coordinate. Find the image of a point under a translation.	What is the distance between (3, 9) and (3, 2)? Where does the point (2, 5) move to after a translation of 3 units right and 4 units down?

The image is congruent to the object after a reflection.	Reflect a shape on a coordinate grid, recognising the image is congruent to the object.	Reflect the shaded shape: **a)** in the y-axis **b)** in the x-axis
The image is congruent to the object after a rotation.	Rotate shapes 90° and 180° around a centre of rotation.	Rotate the shape about the given centre of rotation: **a)** 90° clockwise **b)** 180°
The image is similar to the object after an enlargement.	Enlarge a shape with a given positive scale factor.	Enlarge the following shapes by a scale factor of 3.
	Recognise when one shape is an enlargement of another.	What is the scale factor of the enlargement which transforms shape A onto shape B?

19 Percentages

You will learn how to:

- Recognise percentages of shapes and whole numbers, including percentages less than 1 or greater than 100.

Starting point

Do you remember…

- how to recognise and shade percentages (including 1% and multiples of 5% up to 100%) of shapes?

  For example, shade 84% of a 5 × 5 square grid.

- how to find simple percentages of quantities?

  For example, find 15% of 400.

This will also be helpful when…

- you learn to increase or decrease an amount by a percentage.

19.0 Getting started

Percentages can be greater than 100%

The statements below are about a company called Hats For You.
Group them into possible and impossible statements.

Explain your choices.

a) This year, Hats For You sold 266% as many items as last year.

b) 149% of the employees at Hats For You are women.

c) Hats For You has 135% as many female employees as male employees.

d) 122% of the employees at Hats For You travel less than 50 km to work.

e) In the last year, the number of woollen hats sold by Hats For You decreased by 105%.

f) In the last year, the number of baseball caps sold by Hats For You increased by 105%.

Make up some possible statements using percentages greater than 100%

19.1 Percentages of quantities

Did you know?

In many parts of the world, there are taxes on sales of goods. Shops have to calculate the tax for each product and add it to the cost. When you buy something, part of the cost is tax which goes to the government.

These taxes are given in percentages. For example, in Morocco the percentage is 20% and in Jamaica it is 12.5%

Worked example 1

a) Find 23% of $400.

b) Parvati has $35. Dawid has 120% of this amount. Work out how much money Dawid has.

a) 10% of $400 = $400 ÷ 10 = $40

100% is the whole quantity, or $400

10% is $\frac{10}{100}$ or $\frac{1}{10}$ of $400

This is $400 ÷ 10

100% $400									
10% $40	10% $40	10% $40	10% $40	10% $40	10% $40	10% $40	10% $40	10% $40	10% $40

1% of $400 = $400 ÷ 100 = $4

1% is $\frac{1}{100}$ of $400

This is $400 ÷ 100, or $40 ÷ 10

10% $40									
1% $4	1% $4	1% $4	1% $4	1% $4	1% $4	1% $4	1% $4	1% $4	1% $4

23% of $400 = 2 × $40 + 3 × $4 = $92

23% is 2 × 10% + 3 × 1%

23%		
10% $40	10% $40	1%1%1% $4 $4 $4

Alternative method

Convert 23% to a decimal or fraction:

23% = 0.23 = $\frac{23}{100}$

Then find 0.23 × $400 or $\frac{23}{100}$ × $400

b) 100% of $35 = $35

20% of $35 = $35 ÷ 5 = $7

100% is the whole quantity, or $35

20% is $\frac{20}{100}$ or $\frac{1}{5}$ of $35

This is $35 ÷ 5

100% $35				
20% $7	20% $7	20% $7	20% $7	20% $7

120% of $35 = 100% + 20%

  = $35 + $7 = $42

Dawid has $42

120% is 100% + 20%

120%	
20% $7	100% $35

Alternative method

Convert 120% to a decimal or fraction:

120% = 1.2 = $\frac{120}{100} = \frac{6}{5}$

Find 1.2 × 35 or $\frac{6}{5}$ × 35

Worked example 2

Find $10\frac{1}{2}$% of 7000.

Find of 7000	10% is $\frac{10}{100}$ or $\frac{1}{10}$ of 7000
$\quad$10% of 7000 = 7000 ÷ 10	This is 7000 ÷ 10
$\qquad\qquad$ = 700	

100% 7000									
10% 700	10% 700	10% 700	10% 700	10% 700	10% 700	10% 700	10% 700	10% 700	10% 700

1% of 7000 = 70

$\frac{1}{2}$% of 7000 = 35

1% is $\frac{1}{100}$ of 7000

This is 7000 ÷ 100

$\frac{1}{2}$% is half of 1%

10% 700										
1% 70	1% 70	1% 70	1% 70	1% 70	1% 70	1% 70	1% 70	1% 70	$\frac{1}{2}$% 35	$\frac{1}{2}$% 35

$10\frac{1}{2}$% of \$35 = 10% + $\frac{1}{2}$%

$\qquad\qquad$ = 700 + 35

$\qquad\qquad$ = 735

$10\frac{1}{2}$% is 10% + $\frac{1}{2}$%

$10\frac{1}{2}$%	
10% 700	$\frac{1}{2}$% 35

Tip

Technology question You can use a calculator to calculate a percentage of an amount. For example, to find 23% of 400, you can:
- use the percentage (%) key: type 23% × 400
- use an equivalent decimal: type 0.23 × 400
- use an equivalent fraction: type $\frac{23}{100}$ × 400

Try each method on your calculator.

Exercise 1 $\qquad$ 1-4, 6-12

Tip

In question 1, use your answers to earlier parts to help you with later parts. For example, in part **b)** you can find 90% by subtracting 10% from \$7500.

1 Find the given percentages of \$7500.

a) 10%	b) 90%	c) 50%	d) 20%	e) 1%
f) 2%	g) 99%	h) 5%	i) 15%	j) 3%
k) 18%	l) 23%	m) 91%	n) 56%	o) 78%

2 Find:

a) 45% of 220 $\qquad$ b) 12% of \$50 $\qquad$ c) 6% of 150 $\qquad$ d) 34% of 250 kg

e) 9% of 30 cm $\qquad$ f) 16% of \$75 $\qquad$ g) 24% of 80 $\qquad$ h) 4% of 320

3 Find the given percentages of 4000.

a) 100%	b) 200%	c) 250%	d) 110%	e) 390%
f) 0.5%	g) 2.5%	h) 0.1%	i) 0.4%	j) 5.4%
k) 5%	l) 115%	m) 245%	n) 110.5%	o) 30.3%

4 Find:

a) 1.5% of 200   b) 12.5% of 2000   c) 300% of 24   d) 180% of 15

e) $\frac{1}{2}$% of 800   f) $2\frac{1}{2}$% of 600   g) $\frac{1}{10}$% of 6000   h) $\frac{1}{4}$% of 40 000

5 Use a calculator to find:

a) 28% of 575   b) 67% of 120 kg   c) 11% of $88   d) 125% of 224 km

e) 0.4% of $60   f) 8.6% of 5200 ml   g) 112.5% of 516 cm   h) 822% of 678 m

6 Caleb is calculating 0.5% of 1400 kg.

He says, 'The answer is 0.5 × 1400 = 700 kg.'

Caleb's calculation isn't correct. Explain the error Caleb has made and correct his answer.

7 How many ways can you show that 155% of 300 is 465?

Think about the different ways that you can show your calculation.

8 State which of these are not whole numbers.

a) 1% of 36   b) 25% of 120   c) 200% of 87

d) 150% of 15   e) 0.1% of 600   f) 8.5% of 10 000

g) 5% of 250   h) 101% of 500

> **Tip**
>
> In question 8, you do not have to find the exact result of each calculation.

9 a) 5% of a mass is 6 kg. Find 20% of the mass.

b) 300% of an amount of money is $90. Find 200% of the amount of money.

c) 0.5% of a length is 2 m. Find 10% of the length.

10 a) Find 400% of 15 kg.

b) How many times bigger than 15 is your answer to part a)?

11 An average human baby boy has a mass of 3.5 kg at birth and 10.5 kg at 1 year old.

Which of the statements are correct? Explain your choices.

a) In 1 year, the baby's mass increases by 200%

b) The mass at 1 year is 200% of the mass at birth.

c) In 1 year, the baby's mass increases by 300%

d) The mass at 1 year is 300% of the mass at birth.

12 Write the missing number for each statement below.

a) To find 700% of a quantity, multiply the quantity by ......

b) To find 250% of a quantity, multiply the quantity by ......

c) To increase a quantity by 400%, multiply the quantity by ......

The mass of a newborn kitten is 400 g. When it is a fully grown adult cat, its mass is 4000 g.

d) The cat's final mass as a percentage of its newborn mass is ......%

e) The percentage increase in mass is ....

Thinking and working mathematically activity
True or false? 15% of 80 = 80% of 15. Explore other examples, including percentages greater than 100% (for example, 300% of 50 and 50% of 300), and decimal percentages (for example, 1.5% of 200 and 200% of 1.5).

Is it always true that $x\%$ of $y = y\%$ of x? Try to explain your answer.

Malik wants to calculate 25% of 71% of 400 without using a calculator. He says, 'If I change the order to 71% of 25% of 400, the calculation is easier. But I'm not sure if that gives the same answer.' Does changing the order change the answer? Think about why or why not.

19.2 One quantity as a percentage of another

Worked example 3

a) Express 8 km as a percentage of 20 km.

b) Express shape B's area as a percentage of shape A's area.

A

B

a) The fraction is $\frac{8}{20}$.

As a percentage this is:

$\frac{8\times5}{20\times5} = \frac{40}{100} = 40\%$

Write 8 as a fraction of 20.

Convert to a fraction with denominator 100.

$\frac{1}{20}$

$\frac{8}{20} = \frac{40}{100}$

$\frac{10}{100}$

Calculator method

Calculate $8 \div 20 = 0.4$ and multiply by 100 to convert to a percentage: $0.4 \times 100 = 40\%$

b) B's area is $\frac{17}{10}$ of A's area. As a percentage this is:

$\frac{17\times10}{10\times10} = \frac{170}{100} = 170\%$

Shape A has 10 squares. Shape B has 17 squares.

Write 17 as a fraction of 10, and convert to a percentage.

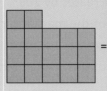
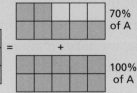

70% of A

100% of A

Worked example 4

Express 5 as a percentage of 16.

$5 \div 16 = 0.3125$	Divide 5 by 16.
$0.3125 \times 100 = 31.25\%$	Multiply by 100 to convert the decimal to a percentage.

1 Find the percentage of each shape that is shaded.

a)   b) c)

Tip

You could write your answer as a fraction first, then convert it to a percentage.

2 Write the first number as a percentage of the second number.

a) 5, 10   b) 5, 20   c) 3, 4   d) 5, 25   e) 4, 5

3 Write down the first quantity as a percentage of the second quantity.

a) $25, $100   b) $60, $100   c) $45, $100   d) 24 cm, 50 cm

e) 15 mm, 50 mm   f) 20 m, 50 m   g) 16 litres, 25 litres   h) 13 litres, 25 litres

i) 28 grams, 50 grams   j) 9 grams, 25 grams   k) $17, $20   l) $23, $25

4 Express shape B's area as a percentage of shape A's area.

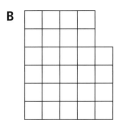

5 Artur says, 'You can't write 80 as a percentage of 20, because 80 is greater than 20.'

Lyudmila says, 'You can write 80 as a percentage of 20. It is $\frac{20}{80}=\frac{1}{4}=25\%$.'

Can you write 80 as a percentage of 20? If so, is the answer 25%?
Explain your answers.

6 Express the first number as a percentage of the second number.

a) 20, 200   b) 30, 200   c) 50, 200   d) 200, 50   e) 15, 30

f) 30, 15   g) 60, 15   h) 2, 400   i) 4, 4000   j) 80, 64

Tip

In question **7e)** and **7f)**, start by writing both quantities in the same units.

7 Use a calculator to express:

a) 128 g as a percentage of 200 g   b) 14 tonnes as a percentage of 25 tonnes

c) 75 kg as a percentage of 20 kg   d) 2 mm as a percentage of 400 mm

e) 213 seconds as a percentage of 5 minutes   f) 42 cm as a percentage of 3 m

8 Mia and Lata are comparing their favourite cricket teams.

Mia says, 'My favourite team is better as they have won 14 matches out of the last 25. They have won more matches.'

Lata says, 'My favourite team is better as they have won 12 matches out of the last 20. They have lost fewer matches.'

Use your knowledge of percentages to explain which team is better.
Give reasons for your answer.

9 **a)** Sebastian buys 250 ml of orange juice. He drinks 75 ml of the juice.
Find what percentage of the juice he drinks.

b) On one day, a café sells 60 cups of coffee. The next day, the café sells 150 cups of coffee.
Write the second day's number of coffees as a percentage of the first day's.

c) Urassaya has a 5 kg bag of rice. One day she cooks and eats 80 g of the rice.
Find the percentage of the rice that is left.

d) Atuat has a piece of rope that is 8 m long. He cuts a 3 m piece of rope.
Find the percentage of the original piece of rope that is left.

10 Isra has 75 ml of water in a container.

The water fills 5% of the total volume of the container.

Find the total volume of the container.

> **Think about**
>
> Patrick says, '180 g is 200% of 60 g, because if you start with 60 g, you can add 60 g twice to make 180 g.' Do you agree with Patrick?

Thinking and working mathematically activity

Below are three puzzles.

1) Two numbers have a difference of 45. One number is 25% of the other number. Find the two numbers.

2) Two numbers have a sum of 120. One number is 150% of the other number. Find the two numbers.

3) Two numbers have a product of 64. One number is 400% of the other number. Find the two numbers.

Explain how you solved each puzzle. Did other students use different methods? What method(s) is/are most efficient?

Consolidation exercise

1 This is a plan of Heidi's garden.

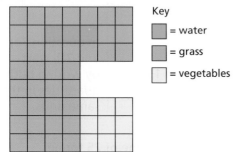

Key
= water
= grass
= vegetables

a) Find the percentage of Heidi's garden that is grass.

b) Find the percentage of Heidi's garden that is used to grow vegetables.

c) Find the percentage of Heidi's garden that is water.

d) Draw a garden plan with 15% taken with water, 40% vegetables and 45% grass.

2 **a)** Li Jing earns $2500 per month and saves 22% of it. Find how much she saves in a year.

b) Each month, Ed earns $2000 and saves $320. Find the percentage of his earnings that he saves.

3 Match each quantity to its equivalent percentage.

48 g out of 60 g	0.1%
$34 out of $40	0.5%
10 m out of 10 000 m	79%
237 people out of 300 people	80%
4 cm out of 800 cm	85%

4 Which would you rather have: 120% of $25 or 25% of $120? Explain your answer.

5 Use a calculator to find the answers.

   a) A jellyfish has a mass of 1500 g.

   95% of its mass is water. Calculate the mass of water in the jellyfish.

   b) An average human adult has a mass of 62 kg. Of this, 37.2 kg is water.

   Calculate the percentage of the average human adult that is water.

   c) A shop sells 120 newspapers one day. The next day, the shop sells 129 newspapers.

   Express the second day's newspaper sales as a percentage of the first day's.

   d) A farmer took 160 cheeses to sell at a market. In the first hour, she sold 7.5% of the cheeses.

   During the rest of the day, she sold 75% of the remaining cheeses.

   Find the number of cheeses she had left at the end of the day.

6 Use the digits 0 – 9 once each to copy and complete the boxes to make the statements true.

   ☐ out of 20 = 3 ☐ %

   ☐☐ out of 75 = 92%

   $\frac{☐☐}{☐☐}$ = 60%

   4 ☐ % of 600 = ☐ 64

End of chapter reflection

You should know that...	You should be able to...	Such as...
Percentages can be greater than 100%	Find a percentage of a quantity.	Find: a) 63% of 800 km b) 250% of $80 c) 0.1% of 3000 people
It can be useful to write a quantity as a percentage of another quantity.	Express a quantity as a percentage of another quantity.	Write: a) 28 g as a percentage of 35 g b) $400 as a percentage of $125

20 Presenting and interpreting data 2

You will learn how to:

- Record, organise and represent categorical, discrete and continuous data. Choose and explain which representation to use in a given situation:
  - waffle diagrams and pie charts
  - line graphs
  - scatter graphs
  - infographics.
- Interpret data, identifying patterns, within and between data sets, to answer statistical questions. Discuss conclusions, considering the sources of variation, including sampling, and check predictions.

Starting point

Do you remember…

- how to draw and interpret simple pie charts?

  For example, what proportion of class 7A have brown eyes?

- how to draw and interpret line diagrams for a single set of data?

  For example, draw a line graph to show the number of students absent each day.

Eye color for students in class 7A

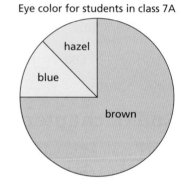

Day	Absences
Monday	11
Tuesday	15
Wednesday	19
Thursday	10
Friday	4

- how to draw a scatter graph and draw on a line of best fit?

  For example, draw a line of best fit on this scatter graph.

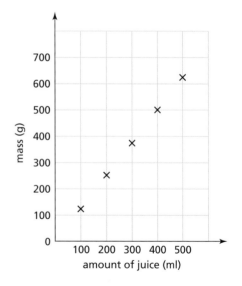

This will also be helpful when…

- you go on to learn about drawing conclusions from diagrams and compare sets of data.

20.0 Getting started

In June 2018, the population of the world was estimated to be 7.6 billion.
The diagram shows some facts about the people that live on this planet.

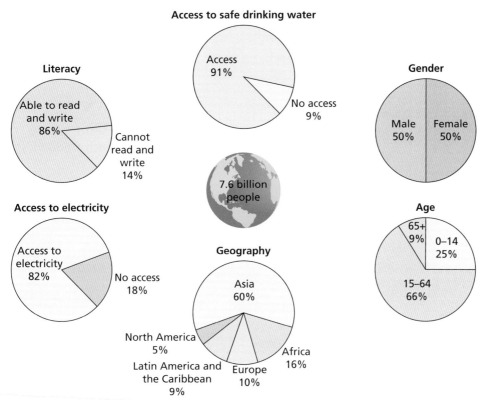

What do the pie charts tell you about the people that live on this planet?

Try to find out some other interesting facts about the people that live on Earth.

20.1 Waffle diagrams and pie charts

Key terms

Waffle diagrams and **pie charts** are used to visualise how a population is made up.
They are diagrams which are used to show proportions.

Waffle diagrams are coloured grids.

Pie charts are circular diagrams – each category of the data is represented by a sector of the circle.

Worked example 1

The table show how 90 students travel to school.
Draw a pie chart to show the information.

Method of travel	Number of students
Walk	15
Car	20
Bus	25
Train	20
Bicycle	10

Walk $\frac{15}{90} \times 360 = 60°$

Car $\frac{20}{90} \times 360 = 80°$

Bus $\frac{25}{90} \times 360 = 100°$

Train $\frac{20}{90} \times 360 = 80°$

Bicycle $\frac{10}{90} \times 360 = 40°$

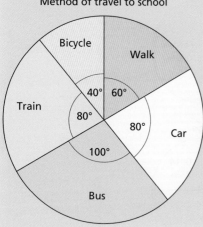

Method of travel to school

Find the fractions of the students who use each method of travel.

Find the angle for each sector by multiplying this by 360°.

Measure the angle of each sector using a protractor.

Remember to label each sector.

Each person in the table is represented by a sector of angle $\frac{360}{90} = 4°$

The angle for each sector can be found by multiply each frequency by 4°.

Tip

Check that your angles add up to 360° before you draw your pie chart.

Worked example 2

This waffle diagram shows the ages of people in a country.

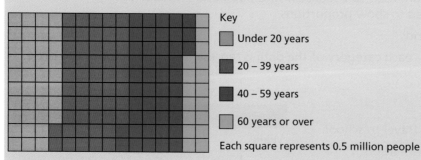

Key

Under 20 years

20 – 39 years

40 – 59 years

60 years or over

Each square represents 0.5 million people

a) What is the total population of this country?

b) Which age group makes up the largest proportion of the country?

c) Show that about 11% of people are aged 60 years or over.

a) Each square represents 0.5 million people. There are 15 × 10 = 150 squares. So the population is 150 × 0.5 = 75 million.	Look at the key to find out how many people each square represents. Find the total number of squares in the waffle diagram. Multiply this by how many people each square represents.	15 squares / 10 squares
b) The 20 – 39 age group makes up the largest proportion of the country.	More squares are coloured red than any other colour. The key shows that red represents the 20 – 39 age group.	38 orange 50 red 45 blue 17 green
c) $\frac{17}{150} \times 100 = 11.3\%$ This is 11% to the nearest whole number.	First find that there are 17 green squares. Then you need to find the percentage that this represents, out of the total of 150 squares.	

Exercise 1

1 A hotel has four types of room. The table shows the number of rooms of each type.

Type of room	Frequency	Angle on a pie chart
Single	11	
Twin	14	
Double	42	
Family	23	

a) Copy and complete the table to show the angle that would represent each type of room on a pie chart.

b) Draw a pie chart to show the information.

2 30 children in a class choose a club to take part in on a Monday lunchtime.
The table shows some information about the choices these children made.

Choice of club	Frequency	Angle on a pie chart
Cookery	7	84°
Choir		108°
Art		72°
Sport		
TOTAL	**30**	**360°**

a) Copy and complete the table.

b) Draw a pie chart to show the information.

3 Here is a breakdown of the ages of the population in India.

Draw a waffle diagram to represent these data.

Use 1 square to represent 10 million people.

> Tip
>
> A waffle diagram does not have to be a
> perfect rectangle in shape.

Age (years)	Population (millions)
0–19	460
20–44	370
45–64	140
65+	50

4 Oti asks 120 students in her school about their favourite subjects.
Her results are shown in the waffle diagram.

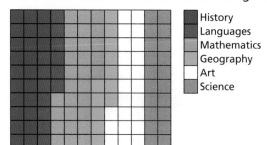

History
Languages
Mathematics
Geography
Art
Science

a) Find how many students each square represents.

b) Find how many students said Geography was their favourite subject.

c) Find what percentage of the students chose Mathematics, to the nearest whole number.

d) Which is more popular, History or Languages? Give a reason for your answer.

5 **a)** Represent the information from question 4 as a pie chart.

b) The school wants to compare the popularity of different subjects.

Explain why a pie chart gives this information more clearly than a waffle diagram.

c) Explain why the waffle diagram is more suitable than a pie chart for finding the number of students in Oti's sample who said each subject.

6 Jacques notes the most common type of clouds in the sky for 130 days.

a) Draw a waffle diagram to represent these data.

b) Explain why the total of 130 means it will be easier to draw a waffle diagram, rather than a pie chart.

Type of cloud	Frequency
Cumulus	26
Cumulonimbus	13
Stratus	46
Cirrus	28
No clouds	17

7 The pie charts show the favourite type of chocolate for 40 boys and 40 girls.

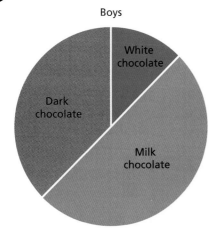
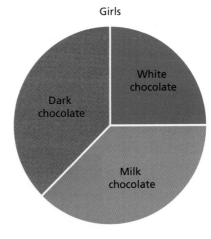

a) How many boys said that milk chocolate was their favourite?

b) How many more girls than boys said they liked white chocolate?

c) Which type of chocolate is favourite for equal numbers of boys and girls?

> **Think about**
>
> If you did not know the total number of boys and the total number of girls would you be able to answer part **c)**?

8 The pie chart shows the different types of vehicle travelling along the road past a school in one day.

a) What proportion of the vehicles were cars?

b) A journalist says 'A quarter of the vehicles on the roads are buses or lorries'.

Do you agree with the journalist? Explain your answer.

c) Marta says 'Half of the people who travel past the school are in a car'.

Do you agree with Marta? Explain your answer.

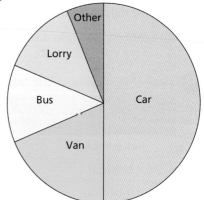

9 Tom investigates the favourite meals for students in his school.

There are 300 boys and 300 girls in the school.

He has this hypothesis:

Pizza is the favourite meal for more boys than girls in the school.

He draws these diagrams to show the data he collects.

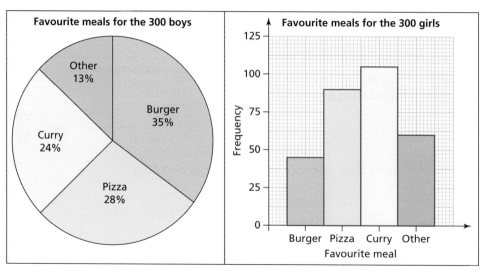

Favourite meals for the 300 boys

Favourite meals for the 300 girls

Comment on his hypothesis. Show how you decided on your answer.

10 A restaurant serves food every evening. The waffle diagram shows the meals ordered over the past two weeks.

a) Find how many meals were ordered.

b) What fraction of the orders were pizza?

c) Find the number of curries they would expect to serve if 200 people come next week.

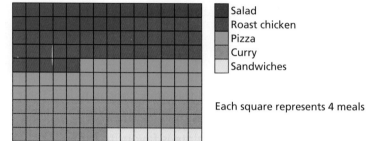

Each square represents 4 meals

The restaurant decides to start serving meals at lunch time.

d) Explain why these data may not be suitable for predicting how many of each meal they will serve.

11 Real data question The table shows how energy is produced in Germany.

Type	Power production in terawatt hours
Renewables	225.7
Nuclear	76
Light Coal	145.5
Hard Coal	83.2
Natural Gas	83.4
Mineral Oil	5.2
Others	27

Source: Clean Energy Wire (CC BY 4.0)

Enter the data in a spreadsheet. Use the spreadsheet to draw a pie chart to represent the data.

Remember to label the sectors and to put a title on your diagram.

12 Ten thousand people watch a football match.

Evie samples 100 people at the match.

She draws this pie chart to show if the people in her sample were children, adults or senior citizens.

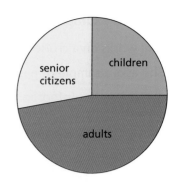

a) Evie concludes that roughly one quarter of all people attending the match are children. Is this conclusion likely to be true or false? Give a reason for your answer.

b) She also concludes that there must be less than 5000 adults watching the match. Give a reason why this conclusion could be false.

Thinking and working mathematically activity

Freya is a nurse. She asks a sample of patients to record the mass of different types of food they ate on one day.

She draws this pie chart to summarise her overall results.

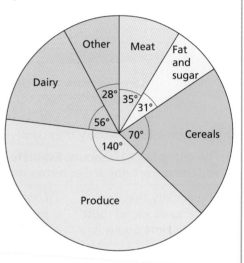

Discuss whether these conclusions are true or false or whether you cannot tell from the pie chart. Give a reason for each answer.

- The mass of cereals eaten on average per day by one of Freya's patients is greater than the mass of meat.

- In America, less than 10% of the food eaten is fat and sugar.

- For a typical patient in the world, less than one third of the food eaten is produce.

- A typical patient ate approximately twice as much dairy as fat and sugar.

Write some conclusions of your own.

20.2 Infographics

Key terms

An **infographic** is a general term for a diagram that is designed to display data in a visually appealing way. They should display the information in a very clear way that makes interpretation quick and easy.

Worked example 3

Mel works on a cruise ship.

She asks a sample of passengers how they spent their time on the cruise.

She draws this infographic to show the average percentage of time that passengers spend on different activities.

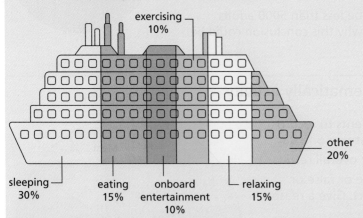

How passengers spend time on the cruise

exercising 10%

other 20%

sleeping 30%

eating 15%

onboard entertainment 10%

relaxing 15%

a) Write down the activity that passengers spend the most time doing.

b) The cruise lasts 120 hours. Estimate the amount of time a typical passenger spends watching the onboard entertainment.

a) The activity that passengers spend most of their time doing is sleeping.	Look for the largest section of the diagram.	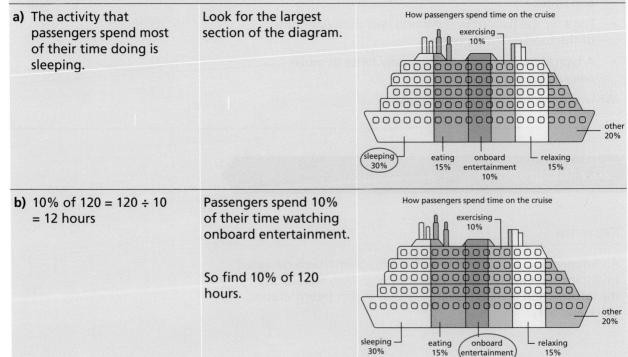
b) 10% of 120 = 120 ÷ 10 = 12 hours	Passengers spend 10% of their time watching onboard entertainment. So find 10% of 120 hours.	

Exercise 2

1 **Real data question** The infographic shows sales of electric cars in some countries in 2017.

Electric car sales by country in 2017

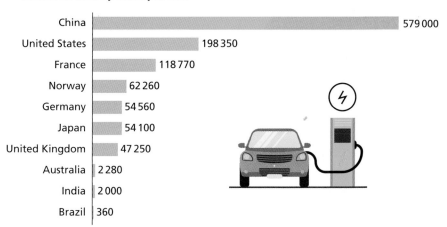

Country	Sales
China	579 000
United States	198 350
France	118 770
Norway	62 260
Germany	54 560
Japan	54 100
United Kingdom	47 250
Australia	2 280
India	2 000
Brazil	360

Source: Global EV Outlook 2018, https://www.iea.org/reports/global-ev-outlook-2018

a) Write down the number of electric cars sold in Germany in 2017.

b) List the countries where more than 100 000 electric cars were sold in 2017.

c) Calculate how many more electric cars were sold in Japan in 2017 than in Australia.

2 A teacher draws this infographic to show how students on a school residential trip spent their time.

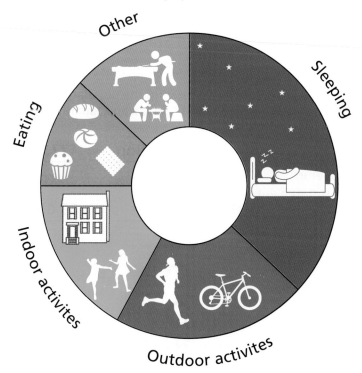

a) What did students spend most of their time doing on the trip?

b) Did students spend more time eating or doing Indoor activities?

3 **Real data question** The infographic shows carbon emissions for the world in 2017.

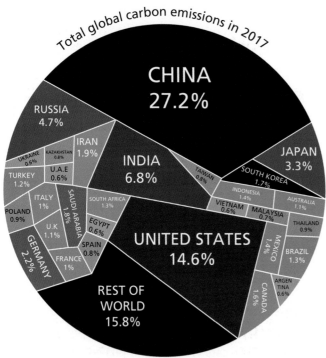

Total global carbon emissions in 2017

CHINA 27.2%

RUSSIA 4.7%

IRAN 1.9%

UKRAINE 0.6% KAZAKHSTAN 0.8%

TURKEY 1.2% U.A.E 0.6%

ITALY 1%

POLAND 0.9%

SAUDI ARABIA 1.8%

SOUTH AFRICA 1.3%

EGYPT 0.6%

GERMANY 2.2%

U.K 1.1%

FRANCE 1%

SPAIN 0.8%

INDIA 6.8%

TAIWAN 0.8%

JAPAN 3.3%

SOUTH KOREA 1.7%

INDONESIA 1.4%

VIETNAM 0.6% MALAYSIA 0.7%

AUSTRALIA 1.1%

THAILAND 0.9%

UNITED STATES 14.6%

MEXICO 1.4%

BRAZIL 1.3%

CANADA 1.6%

ARGENTINA 0.6%

REST OF WORLD 15.8%

Source: BP Statistical Review of World Energy 2020

Discuss

Do you think that the way the data is being presented in this infographic is effective? Is it easy to interpret? Do the percentages match the areas of each section? Can you suggest a better way to present the data?

a) Write down the percentage of the world's carbon emissions produced by Japan in 2017.

b) How do the emissions of Turkey compare with the emissions of Canada?

c) Danesh says the top three countries emit more than half the world's carbon. Is he correct? Justify your answer.

4 **Real data question** The infographic shows information about the population of Japan and Canada.

a) Write down the total population of Canada.

b) Mo says that, in 2019, a greater percentage of the population was aged 65 and older in Canada than in Japan. Comment on his statement.

c) Compare the life expectancies of people living in Canada and Japan.

d) Comment on how the percentage of the population in Canada that is aged 65 and older is forecast to change between 2019 and 2030.

Comparing the populations of Canada and Japan

Total population

Canada **37.0 million**   Japan **127.2 million**

Life expectancy

Canada **83.0 years**   Japan **85.0 years**

Population aged 65 and over

Canada 2019 **17.6%**   Canada 2019 **28.0%**

Canada 2030 **22.8%**   Canada 2030 **30.9%**

Source: Copyright © 2019 by United Nations, made available under a Creative Commons license (CC BY 3.0 IGO) http://creativecommons.org/licenses/by/3.0/igl/

5 Real data question This infographic shows the percentage of the population that used the internet in 2018 for 10 countries.

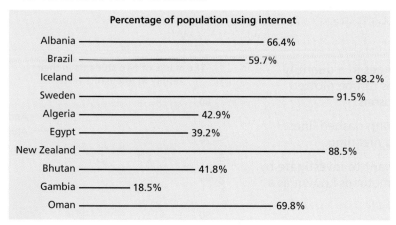

Source: UN Data

a) Which country has the highest percentage of its population using the internet?

b) Write down the percentage of the population of Egypt that used the internet in 2018.

c) Compare the percentages of the population using the internet in Brazil and Oman.

Real data question **Thinking and working mathematically activity**

The table shows the water used per person each day in 10 countries.

Create your own infographic to display the data.
Think about:

- what your title will be
- what theme you will use for the design
- what colours you will use to ensure that the data is still clear
- how big text needs to be to clearly display your data
- how you can make your design eye catching
- how you can use a suitable background to highlight any key themes surrounding water use.

Write some conclusions from your diagram.

Look at another student's infographic. Think about:

- what parts of their design help to show any patterns in the data clearly

Country	Water used each day per person (litres)
Turkmenistan	16 281
Chile	5 935
Guyana	5 283
Uzbekistan	4 754
Tajikistan	4 547
Kyrgyzstan	4 281
United States	3 794
Iran	3 707
Estonia	3 580
Azerbaijan	3 556

Source: Worldometers.info

- whether there are things about their design that make it difficult to interpret the data
- what advice you would give them to help improve their design.

Key terms

Line graphs are used to show changes in a quantity over time. Two or more line graphs can be plotted on the same axes to make comparisons.

Sometimes points are connected with dashed lines when intermediate values have no meaning.

A **hypothesis** is an idea that you want to investigate by collecting data. A hypothesis is sometimes known as a conjecture.

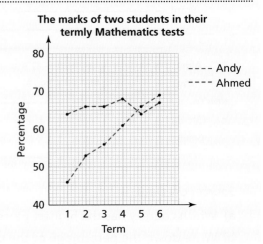

The marks of two students in their termly Mathematics tests

Worked example 4

The graph shows the heights of Alishbah and Ben on their birthdays.

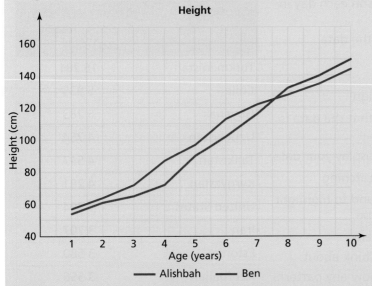

a) When was the first birthday that Alishbah was taller than Ben?

b) On which birthday is Ben the same height as Alishbah was on her 4th birthday?

c) Show that Ben's height more than doubled between his 1st and 10th birthdays.

a) The first birthday that Alishbah was taller than Ben was her 8th birthday.	Alishbah is taller than Ben for the first time when the blue line is above the red line.	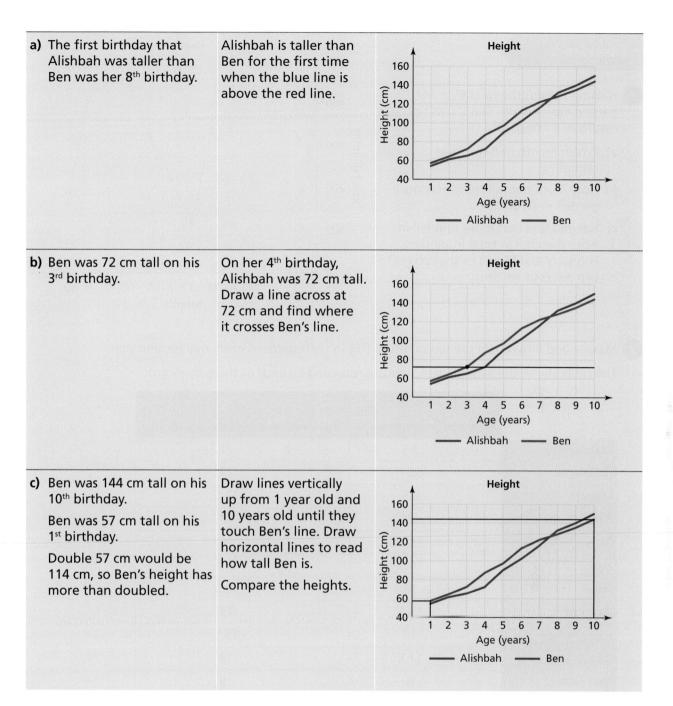
b) Ben was 72 cm tall on his 3rd birthday.	On her 4th birthday, Alishbah was 72 cm tall. Draw a line across at 72 cm and find where it crosses Ben's line.	
c) Ben was 144 cm tall on his 10th birthday. Ben was 57 cm tall on his 1st birthday. Double 57 cm would be 114 cm, so Ben's height has more than doubled.	Draw lines vertically up from 1 year old and 10 years old until they touch Ben's line. Draw horizontal lines to read how tall Ben is. Compare the heights.	

Exercise 3

1 Hakima records the rainfall in her garden each month. She shows her results in a line graph.

a) Which month had the most rainfall?

b) How much rainfall fell in Hakima's garden in August?

c) Hakima says that more rain fell in April than fell in total in January, February and March. Is she correct? Explain your answer.

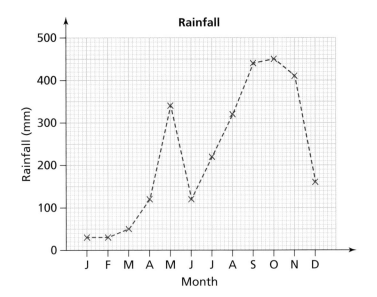

2 Maxine and Rita record the temperature (°C) in their gardens every day for one year.

The table shows the average temperature recorded by each of them by month.

	Maxine's data	Rita's data
	Average temperature (°C)	Average temperature (°C)
January	3.5	21
February	5	23
March	6.5	22.5
April	8	18
May	11.5	16
June	16.5	13.5
July	18	12.5
August	18.5	13
September	16	15.5
October	11.5	18
November	8	19
December	5.5	20.5

a) Plot both sets of data as line graphs on one set of axes.

b) Use your graph to write down which month saw the greatest different in temperatures recorded by the two girls. What was the difference in temperatures for that month?

c) Use your graph to find how many months the average temperature in Rita's garden was greater than in Maxine's.

3 Emanuel owns a small gym. Emanuel makes this claim:

'The number of members at the gym more than doubled between 2012 and 2018'.

He draws this line graph to show how the number of members at the gym has changed.

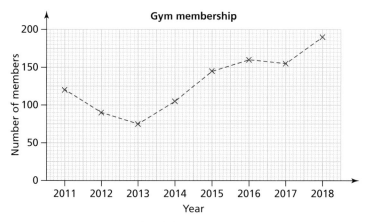

Comment on Emanuel's claim.

4 Two museums record the number of visitors they have over a one-week period.

a) How many visitors did Museum A have on Tuesday?

b) Describe any patterns you see across the week.

c) Show that, overall, Museum B had more visitors than Museum A.

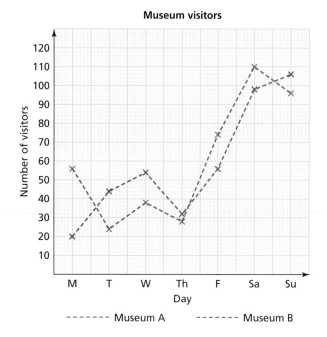

5 A scientist measures the height of three seedlings for eight days.

a) Write down the height of Seedling C on Day 5.

b) Find the change in the height of Seedling B between Day 4 and Day 6.

c) Which seedling was tallest on Day 3?

d) Which seedling grew by the least over these eight days?

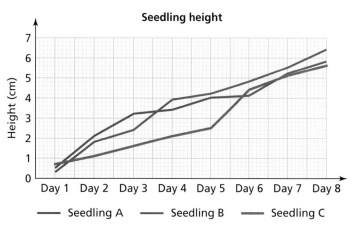

6 The table shows the temperature of hot water in two different flasks for 25 hours.

Time (hours)	0	5	10	15	20	25
Flask A temperature (°C)	100	79	61	47	37	30
Flask B temperature (°C)	100	70	49	34	24	17

a) Draw a line graph to show both these sets of data.

b) Use your line graph to estimate the temperature in the two flasks after 8 hours.

c) Find the difference in the temperatures at 18 hours.

d) Which flask is better insulated? Explain how you know.

Thinking and working mathematically activity

Flynn is choosing between booking a holiday in April or in August. He is also choosing between two holiday resorts.

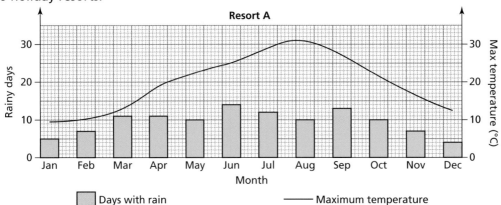

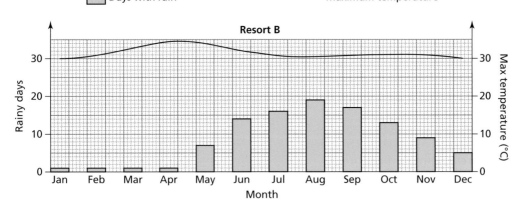

- Make some conclusions about the weather at these two resorts.

Flynn likes hot weather and wants a low chance of rain.

- Recommend which month he should go on holiday and which of the resorts he should choose. Explain your choices.

20.4 Scatter graphs

Key terms

A **scatter diagram** shows whether or not two variables are related. One variable is plotted on the horizontal axix and the other is plotted in the vertical axis.

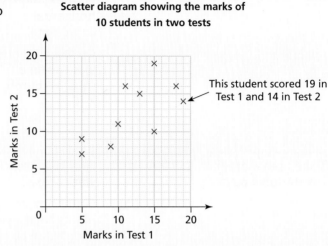

Scatter diagram showing the marks of 10 students in two tests

This student scored 19 in Test 1 and 14 in Test 2

Worked example 5

The scatter graph shows the amount of water (in litres) that Sophie uses on ten different days. The amount of water is plotted against the maximum temperature (°C) on these days.

a) Write down the highest temperature recorded on the chart.

b) Count how many days Sophie used more than 150 litres of water.

c) How is Sophie's use of water related to the temperature?

d) Draw a line of best fit on the graph.

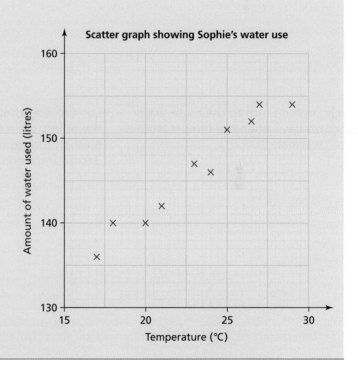

Scatter graph showing Sophie's water use

a) The highest temperature recorded is 29 °C.	Temperature is plotted on the horizontal axis. Find the point with the highest horizontal coordinate. On the horizontal axis two small squares represent 1 °C.	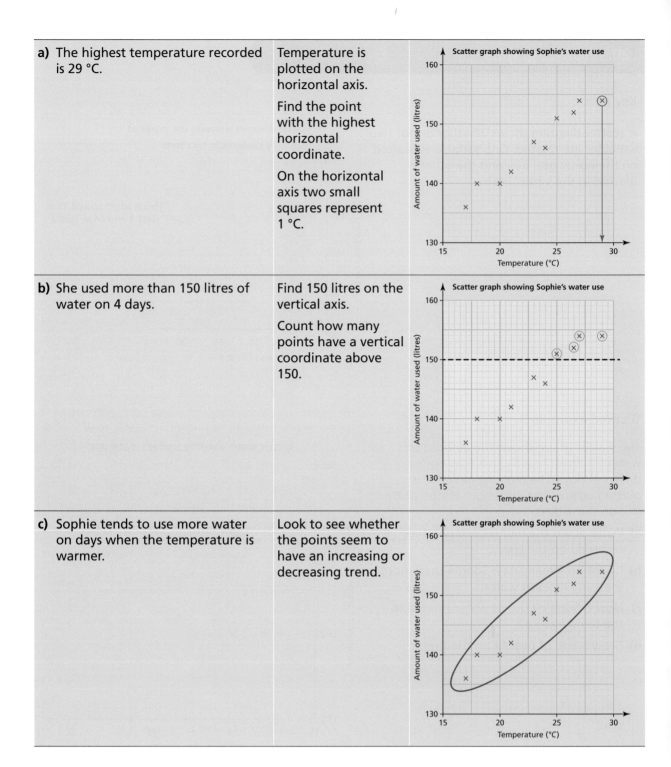
b) She used more than 150 litres of water on 4 days.	Find 150 litres on the vertical axis. Count how many points have a vertical coordinate above 150.	
c) Sophie tends to use more water on days when the temperature is warmer.	Look to see whether the points seem to have an increasing or decreasing trend.	

d)

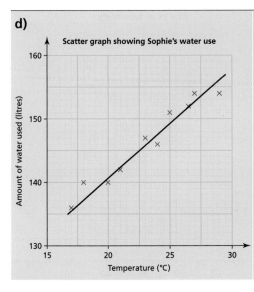

Scatter graph showing Sophie's water use

The line of best fit should follow the trend of the points.

It should have approximately the same number of points on either side of it.

It does not need to go through any of the points.

Exercise 4

1 Draw scatter graphs to show each set of data.

a)
Length of parcel (cm)	16	20	20	24	26	30	32	35
Mass of parcel (g)	340	300	460	320	570	680	640	650

b)
Time spent on homework (min)	10	25	40	45	50	60	65	80
Time spent on computer games (min)	90	100	75	50	60	25	0	40

2 Preeti measures the length and width of some leaves.

Length of leaf (cm)	7.6	6.5	8.9	6.3	5.4	8.2	9.6
Width of leaf (cm)	4.5	4.1	5.2	3.7	3.5	4.9	5.6

a) Draw a scatter graph to show Preeti's data.

b) Draw a line of best fit on your scatter graph.

c) How is the width of a leaf related to its length?

d) Use your line to predict the width of a leaf that has length 8.4 cm.

3 The scatter graph shows test marks for a group of students in Maths and Science.

a) How many results are recorded on the scatter graph?

b) Karla scored the highest mark in the Science test. What was her mark on the Maths test?

c) Samira scored the same mark on both tests. What was this mark?

d) How many students scored more than 25 marks in Maths?

e) Delete the incorrect word in the sentence below to make it correct.

Students who did well in the Science test tended to do better/worse in Maths than students who did less well in Science.

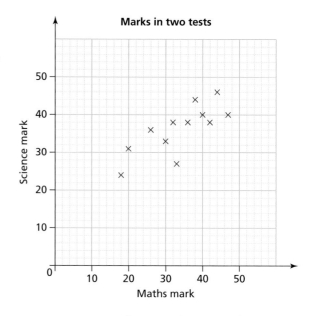

Marks in two tests

4 The scatter diagram shows the mass of sugar and the mass of flour in some cake recipes.

a) Write down how many cakes are represented on the scatter graph.

b) One of the cakes contains 300 grams of flour. Write down the mass of sugar in this cake.

c) What is the relationship between the mass of sugar and the mass of flour in a cake?

d) Greg makes a similar cake to those shown on the scatter graph. He plans to use 150 grams of sugar and 340 grams of flour.

Explain why Greg may have made a mistake with his recipe.

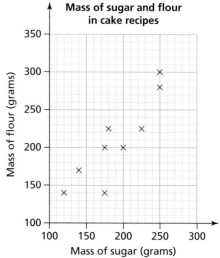

Mass of sugar and flour in cake recipes

5 Emily has gathered data from her class to see if there is a relationship between hand span and foot length.

a) Plot a scatter graph of these data.

b) Draw a line of best fit on your scatter graph.

c) Describe the relationship between hand span and foot length.

Hand span (cm)	Foot length (cm)
22.6	24.4
23.2	24.2
23.9	24.1
19.5	20.7
17.8	19.1
16.4	17.1
22	23.4
17.3	18.3
16.9	17.9

6 Jayden has the following hypothesis.

The warmer the weather is, the fewer customers I have in my café.

He gathers data for nine days.

Temperature (°C)	16	23.1	25	23.5	19.4	16.4	20.5	17.7	18.6
Number of customers	114	28	23	36	77	94	52	82	85

a) Plot the graph of his data.

b) Draw a line of best fit on your graph.

c) Is Jayden's hypothesis correct? Justify your answer.

d) The next day it is predicted to be 20 °C. Use your line of best fit to predict, approximately, how many customers he will have.

7 Tyrone has this hypothesis:

The heavier the mass of an animal the slower it is able to run.

He collects this data for 12 animals.

Animal	Mass (kg)	Running speed (km/h)
Antelope	37	105
Black bear	135	48
Capra Goat	70	45
Fox	6	72
Gazelle	30	97
Hyena	45	50
Llama	72	56
Panther	150	59
Reindeer	120	80
Sheep	65	60
Springbok	34	97
Wolf	40	64

a) Draw a scatter graph and add a line of best fit.

b) What does your line of best fit tell you about Tyrone's hypothesis?

c) Use your line of best fit to predict the running speed of an animal with a mass of 80 kg.

d) What data could you include to investigate further?

8 Ten students take a music qualification.

The students need to take a theory and a practical examination.

The scatter graph shows the percentage marks for the two exams.

a) Write down the lowest percentage mark on the practical examination.

b) The theory examination is marked out of 50.

The practical examination is marked out of 100.

To pass the qualification, students must score a total mark of at least 108 marks.

Catriona was one of these 10 students.

She scored 82% on the practical examination.

Did she pass the qualification? Explain your answer.

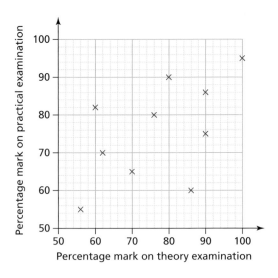

▼ **Thinking and working mathematically activity**

Use the internet to collect some numerical data about 10 vehicles (for example, cars, planes or ships). Your data could be about mass, speed, engine size, fuel efficiency, etc.

Choose two of your variables and make a prediction as to what kind of relationship, if any, there will be.

Enter your data on a spreadsheet and use it to draw a scatter graph.

Add a line of best fit, if appropriate, to your graph.

Use your graph to test your prediction.

Tip

Think about:

- what axis scales you will use to show any trends clearly
- whether your axes need to start at zero
- what axis labels, and units, and title you will put on your graph.

Consolidation exercise

1 **Real data question** The graph shows the average age at first marriage for men and women in Scotland.

a) Write down the age for males at first marriage in Scotland in 1975.

b) Fred says that the differences between the ages of men and women, when they first marry, is lower in 2017 than in 1975. Is he correct? Give a reason for your answer.

The average age at first marriage in Scotland increased in 2017 for both males and females

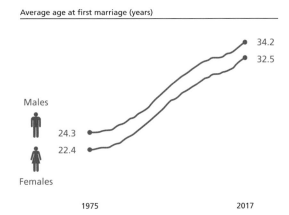

Source: National Records of Scotland © Crown copyright and database right 2018.

---

2 Fatima investigates how quickly water cools in two different containers.

Container 1

Time (min)	0	5	10	15	20	25	30
Temperature (°C)	98	73	66	60	55	51	48

Container 2

Time (min)	0	5	10	15	20	25	30
Temperature (°C)	98	68	60	53	47	42	38

a) Draw a line graph to show Fatima's results. Plot both sets of results on the same axes.

b) In which container did the water cool most quickly?

c) Estimate the temperature of the water in Container 2 after seven minutes.

3 90 students in a school each choose one of these subjects to study in next year: Art, Music, Cookery or Dance.

The pie chart shows their choices.

Explain whether each of these statements is true or false.

a) The angle for cookery on the pie chart is 64°

b) More people chose Music than Dance.

c) More than twice as many people choose Art over Music.

d) 21 people chose Dance.

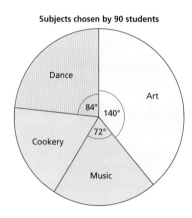

Subjects chosen by 90 students

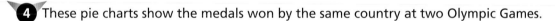

4 These pie charts show the medals won by the same country at two Olympic Games.

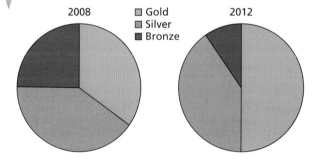

Vincent says, 'The team did better in 2012 than in 2008 because they won more gold and silver medals.' Do you agree with Vincent? Explain your answer.

5 Harry wants to know if the population of the city where he lives has tripled between 1915 and 2015. He sees this line graph on the internet showing the city's population.

Use the information in the line graph to make a conclusion about whether or not the population of the city has tripled. Show how you reached your conclusion.

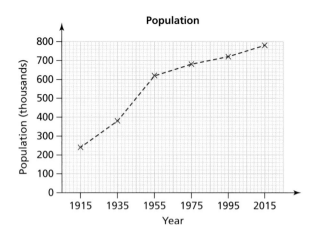

6 Flo records the number of telephone calls she receives in the morning and in the afternoon on ten different days.

She shows her results on a scatter graph.

Write down if each of these statements is true or false.

a) On the day that Flo received 15 calls in the morning, she received 14 calls in the afternoon.

b) On three days Flo received fewer than 16 telephone calls in both the morning and the afternoon.

c) The most number of calls she received in an afternoon was 19.

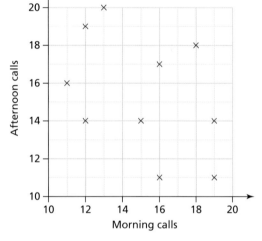

d) The largest total number of calls Flo received in one day was 36.

7 The waffle diagram shows the composition of the gases in the atmosphere.

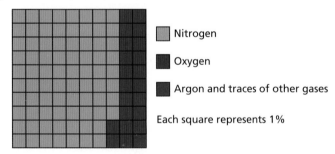

□ Nitrogen

■ Oxygen

■ Argon and traces of other gases

Each square represents 1%

a) Write down the percentage of the atmosphere that is nitrogen.

b) Sarah says that the percentage of nitrogen in the atmosphere is just under four times the percentage of oxygen. Is she correct? Give a reason for your answer.

End of chapter reflection

You should know that...	You should be able to...	Such as...
Pie charts and **waffle diagrams** show proportions.	Draw and interpret pie charts and waffle diagrams	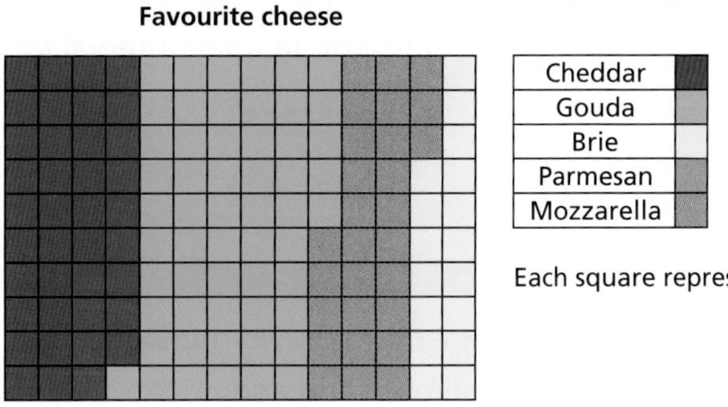Favourite cheese Cheddar Gouda Brie Parmesan Mozzarella Each square represents 50 people Find the percentage of people who preferred Brie.
Infographics are used to represent data in pictorial form.	Read data to draw conclusions from an infographic. Draw your own infographic to present data.	Research the different ways that people in your class travel to school. Design an infographic to present these data.
A scatter graph shows whether or not two variables are related.	Draw and interpret a scatter graph.	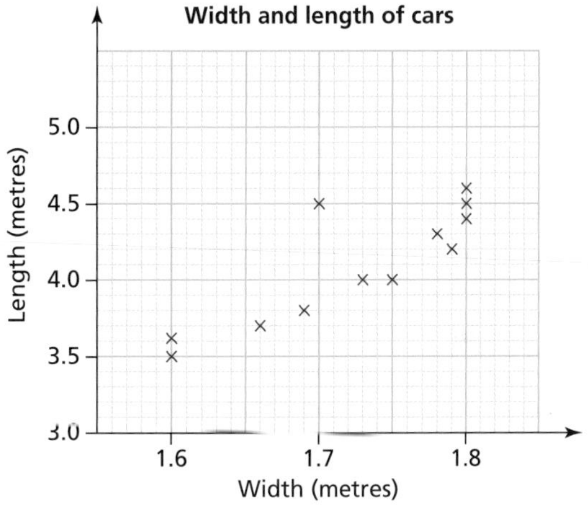Width and length of cars Length (metres) 5.0 4.5 4.0 3.5 3.0 1.6 1.7 1.8 Width (metres) Find how many of the cars have a length less than 4 m.
A line graph can show several sets of data plotted against time.	Draw or interpret line graphs interpreting two or more sets of data.	Sam and Chuka take a maths test every month. 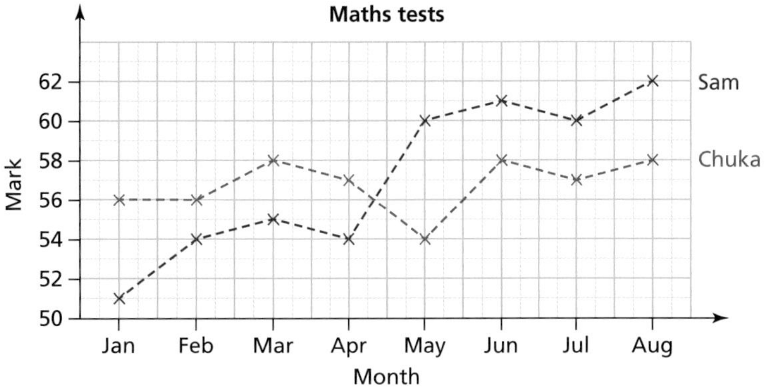Maths tests Mark 62 60 58 56 54 52 50 Jan Feb Mar Apr May Jun Jul Aug Month Sam Chuka Find the difference between the marks that Sam and Chuka scored in May.

21 Equations and inequalities

You will learn how to:

- Understand that a situation can be represented either in words or as an equation. Move between the two representations and solve the equation (integer coefficients, unknown on one side).
- Understand that letters can represent an open interval (one term).

Starting point

Do you remember…

- how to use letters to represent unknown numbers?

  For example, doubling a number and adding 3 could be represented by the expression $2x + 3$.

- what a term, a coefficient, an expression and an equation are?

  For example, in the expression $2x + 3$, the coefficient of x is 2.

- the correct order of operations used to calculate with numbers and algebra?

  For example, with $2x + 3$, the multiplication by 2 is done before adding the 3.

This will also be helpful when…

- you learn about using and interpreting graphs.

21.0 Getting started

Owen orders some pizzas. The cost of each pizza is $4 and the delivery charge is $3.

If $C is the total cost, and p is the number of pizzas he orders, he can write a formula for this:

$$C = 4p + 3$$

What is the total cost for three pizzas?

Owen spent $27. How many pizzas did he order?

Show a partner how you found your answer.

Owen and his friends have $50 to spend on pizzas. How many pizzas can they order?

Show a partner how you found your answer.

Another pizza company does not charge for delivering pizzas, but their pizzas cost a bit more, $4.50. Investigate which company you would order from if you wanted 3 or 4 pizzas. How about 7 or 8 pizzas? For how many pizzas would the two companies' charges be the same?

21.1 Forming and solving equations

Key terms

Sometimes you use a letter to represent an unknown quantity or value that you want to find.

You can find the quantity or value that the letter represents when you have an equation.

An **equation** is a mathematical statement that includes an equals sign (=).

For example, if you know that when you double a number and add 7, you get the answer 19, you could write this as the equation:

$2x + 7 = 19$

Solving an equation means finding the value of the unknown quantity that makes the equation true.

If the equation is $2x + 7 = 19$, then solving it means finding the value of x on its own that makes the equation mathematically correct.

To keep an equation balanced you must always do exactly the same to both sides of the equation.

To 'undo' an operation you need to do the inverse operation.

Worked example 1

Solve these equations:

a) $3x = 12$ b) $2x + 3 = 17$ c) $20 = 3x - 1$

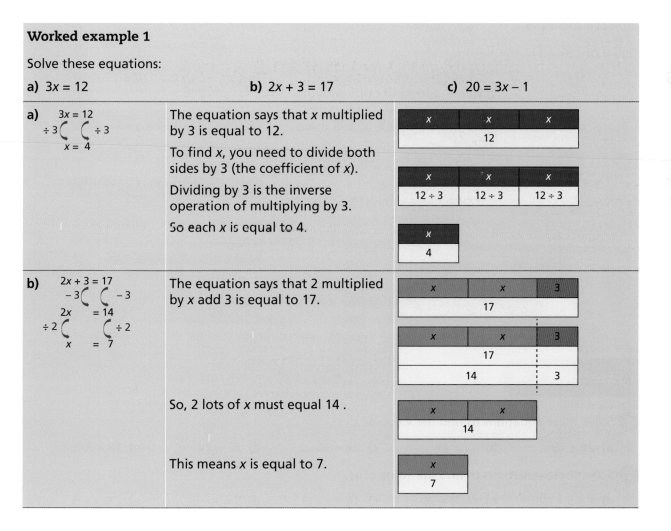

c)
$$20 = 3x - 1$$
$$+1 \quad +1$$
$$21 = 3x$$
$$\div 3 \quad \div 3$$
$$7 = x$$

The equation says that 20 is equal to x multiplied by 3 subtract 1.

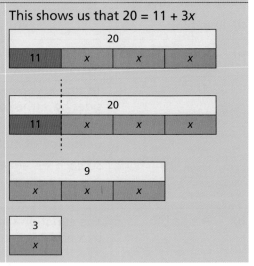

So x multiplied by 3 must be equal to 21.

This means x is equal to 7.

Tip

You can check that you have the correct value for x by putting that value of x back into the equation. In part **c)** of Worked example 1 you saw that if $20 = 3x - 1$ then $x = 7$. To check you need to see if $20 = 3 \times 7 - 1$, which it does.

Think about

Why do you subtract the 3 before dividing by 2 in part **b)** above?

Does the order of arithmetic operations matter?

Worked example 2

Solve $20 - 3x = 11$

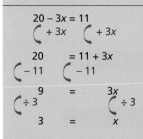

To make the unknown amount positive, add $3x$ to both sides of the equation, then the equation becomes '20 equals 11 add three times x'.

So x multiplied by 3 must be equal to 9...

...which means x must be equal to 3.

This shows us that $20 = 11 + 3x$

Exercise 1

1 Solve these equations to find the value of x:

    a) $2x = 16$ **b)** $3x = 21$ **c)** $9x = 36$ **d)** $22 = 2x$ **e)** $18 = 3x$

2 Solve these equations to find the value of x:

    a) $x + 5 = 11$ **b)** $x - 2 = 20$ **c)** $14 = x + 3$ **d)** $8 = x - 5$ **e)** $2 = x - 3$

3 Solve these equations to find the value of x:

a) $2x + 3 = 11$ b) $3x + 9 = 21$ c) $4x + 7 = 27$ d) $2 + 6x = 26$

e) $5 + 3x = 14$ f) $20 = 3x + 2$ g) $31 = 7x + 3$ h) $38 = 3x + 14$

4 Solve these equations to find the value of x:

a) $2x - 3 = 13$ b) $6x - 9 = 21$ c) $5x - 2 = 23$ d) $7x - 1 = 27$

e) $4x - 3 = 17$ f) $25 = 8x - 7$ g) $21 = 3x - 6$ h) $34 = 7x - 8$

5 Solve these equations to find the value of the unknown:

a) $12 - x = 4$ b) $14 = 19 - x$ c) $11 - 2x = 7$ d) $18 - 4x = 6$

e) $25 - 3x = 16$ f) $32 - 5x = 7$ g) $40 - 6x = 4$ h) $17 = 31 - 2x$

6 Which is the odd one out? Explain your answer.

A: B: C: D: E:

$3x - 4 = 23$ $1 + 2x = 19$ $40 = 6x - 8$ $7x = 63$ $39 = 4x + 3$

7 Pamela is solving the equation $2x + 1 = 33$

Here is her working out:

$2x + 1 = 33$

$3x = 33$

$x = 11$

Explain what mistake Pamela has made, and correct her solution.

8 Quentin is solving the equation $2x - 7 = 17$

Here is his working out:

$2x - 7 = 17$

$2x = 10$

$x = 5$

Explain what mistake Quentin has made, and correct his solution.

9 Match each expression to the correct statement.

$3x - 9$ Subtract 3 times a number from 9

$3 + 9x$ Divide a number by 3 then add 9

$9 - 3x$ Multiply a number by 3 then subtract 9

$9 + \frac{x}{3}$ Divide a number by 3 then subtract 9

$\frac{x}{3} - 9$ Multiply a number by 9 then add 3

10 The blue shape here has a perimeter of 42 cm.

a) Write an expression, in terms of x, for the perimeter of this shape.

b) Use this expression to write an equation for the perimeter.

c) Solve this equation to find the value of x.

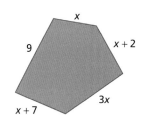

11 Ellen thinks of a number, x. She multiplies her number by 3 and then adds 2. She gets an answer of 23.

    a) Write an equation in x to represent this number puzzle.

    b) Solve your equation to find x.

12 Pedro has 4 identical packets of biscuits. Each packet contains x biscuits. One packet has been opened and 3 biscuits have been removed.

    a) Write an expression using x for the number of biscuits that Pedro has in total.

    Pedro has 53 biscuits.

    b) Write an equation in x to show this information.

    c) Solve your equation to find x, the number of biscuits in a packet.

13 Ajay's mother is 42. Ajay notices that if he triples his age and subtracts 3, the answer is his mother's age. Let x be Ajay's age.

    Write and solve an equation to find the value of x.

 Thinking and working mathematically activity

Start with $x = 2$.
What equations can you make with this being the answer?

Think about different arithmetic operations you can use in your equation, and make at least five different equations that all have the answer $x = 2$.

Now choose another value for x and write some equations for that number.

Challenge a partner to solve your equations to find your value for x.

21.2 Inequalities

Key terms

You can use an **inequality** when you want to describe a situation mathematically, but do not know an exact answer. The inequality describes the set of numbers that could all fit the given situation.

The symbol '>' is used for **greater than** statements, the symbol '<' is used for **less than** statements.

For example, '3 is less than 5' is written as $3 < 5$.

We can also use a letter to represent an unknown number in an inequality.

For example, suppose Lisa is 8 years old and we know that Maria is younger than Lisa. We could use m to represent Maria's age and write $m < 8$.

An inequality can also be represented on a number line, for example for $m < 8$:

The open circle at the end of the arrow shows that you go right up to the 8, but do not include it.

Worked example 3

n is a number that is less than 7.

a) Write an inequality statement for n.

b) Show the possible values of n on a number line.

c) If n is an integer, what is the largest value it can be?

a) $n < 7$	n is less than 7, so the symbol $<$ replaces the words 'is less than'
b) 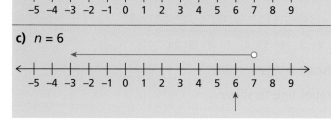	n is less than 7, so you need to show the possible values going down from 7, but not including 7.
c) $n = 6$ 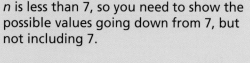	An integer is a positive or negative whole number. The largest whole number less than 7 is 6.

> **Did you know?**
>
> The greater than ($>$) and less than ($<$) symbols were introduced in 1631 in a book by British mathematician Thomas Harriot. However, the symbols were actually invented by the book's editor. Harriot had used triangular symbols, but the editor modified them to resemble the modern less than and greater than symbols.

Exercise 2

> **Tip**
>
> You can remember which inequality symbol to use as the pointed end points to the smaller number and the bigger end to the bigger number.

1 Write these as inequalities.

   a) m is less than 5   b) t is more than 10

   c) x is more than -2 d) f is less than 0

2 Draw number lines to represent the inequalities in question 1.

3 Write the correct inequality signs in the gaps.

   a) 9 _____ 12   b) 271 _____ 217   c) -8 _____ 4   d) -1 _____ -4

   e) $\frac{3}{5}$ _____ $\frac{2}{5}$ f) 0.4 _____ $\frac{1}{2}$ g) 5.3 _____ 5.6   h) 0.1 _____ 0.07

4 Match the inequalities to statements that could describe them.

$x > 4$	There are fewer than 4 people in the car.
$x < 4$	It is warmer than $-4\ °C$ today.
$x > -4$	It is colder than $-4\ °C$ today.
$x < -4$	There are more than 4 people on the bus.

5 Match the inequality statements to the number lines that show them.

a) $x > 3$ b) $x < 6$ c) $x < -3$ d) $x > 6$

(i)

(ii)

(iii)

(iv)

6 Amelie is x years old, Elijah is y years old and Noah is 7 years old.

Write an inequality statement and draw a number line to show:

a) Amelie is older than Noah.

b) Elijah is younger than Noah.

c) Who is the youngest? Explain how you know.

7 n is a number such that $n > 3$. Choose possible values of n from this list.

-3 0   5.4   3.003   1.56 $7\frac{1}{5}$

8 p is a prime number and $p > 15$. Write down two possible values for p.

9 a) Sort these numbers into two groups according to the inequalities.

2	-4.5	$\frac{2}{5}$	3.05
12	150	6.5	-3
-12	0	15.65	7

$\boxed{n < 3}$ $\boxed{n > 7}$

b) Are there any numbers that are not in either group?
If so, how could you describe this group of numbers?

> **Thinking and working mathematically activity**
>
> A box contains some socks that are either coloured red or blue or green.
>
> The number of red socks is more than 3.
>
> The number of blue socks is more than 4.
>
> The number of green socks is more than 1.
>
> The total number of socks is less than 14.
>
> Find possible values for the number of red, blue and green socks in the box.
>
> How many possible answers are there?
>
> Can you convince a friend that you have found all possible answers?

> **Discuss**
>
> For $x > 5$, what is the largest number you could have?
>
> For $y < 4$, what is the smallest number you could have?

Consolidation exercise

1 Solve these equations to find the value of x:

a) $5x = 15$ **b)** $x + 12 = 20$ **c)** $4x + 1 = 21$

d) $7x - 4 = 38$ **e)** $18 = 3x - 6$ **f)** $100 = 3x + 25$

2 Solve these equations to find the value of the unknown:

a) $13 = 4x - 3$ **b)** $12 - 2x = 2$ **c)** $5 = 40 - 5x$

3 Sort these equations into two groups according to the value of x.

a) $x - 7 = 2$ **b)** $7 - x = 1$

c) $2x + 7 = 25$ **d)** $4x - 5 = 7$

e) $9 - 2x = 1$ **f)** $24 - 2x = 4$

g) $36 = 4x + 12$ **h)** $8 = 5x - 32$

 $x < 5$ $x > 5$

4 The perimeter of this blue rectangle is 30 cm.

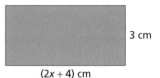

a) Write an expression for the perimeter of the rectangle.

b) Use your expression to write an equation for the perimeter of the rectangle.

c) Solve your equation to find the value of x.

3 cm

$(2x + 4)$ cm

5 Kelly and Emma are solving the equation $2x - 9 = 13$.

Kelly says that the answer is $x = 2$. Emma says that the answer is $x = 11$.

Who is correct?

Can you explain what mistake the other person made?

6 Write each of these statements as an inequality. Show each inequality on a number line.

a) d is less than 4   **b)** n is more than 2   **c)** p is more than -2

7 n is a number such that $n < 7$. Write down which of these numbers could be possible values for n:

5.34   12 -13.7 $7\frac{1}{2}$ 6.99   7

8 I think of an integer, n.

a) My number is less than 12. Write an inequality to show this.

b) My number is more than 8. Write an inequality to show this.

c) Draw both of these inequalities on a number line.

d) What are the possible values of my number?

End of chapter reflection

You should know that...	You should be able to...	Such as...
An equation is a mathematical statement containing an equals sign and an unknown variable. You can solve an equation to find the value of the unknown amount by using inverse operations on both sides of the equation.	Solve an equation to find the value of its unknown term.	Solve: **a)** $3x = 27$ **b)** $21 = 2x - 1$ **c)** $25 - 4x = 17$
	Construct an equation to solve a problem.	The perimeter of this rectangle is 34 cm. Write and solve an equation to find the width and length. x [rectangle] $x + 5$
An inequality represents a set of possible solutions in a given situation. It can be represented on a number line.	Write an inequality that describes a situation and represent it on a number line.	The number of people, p, at a bus stop is less than 6. $p < 6$ 0 1 2 3 4 5 6 7 8 9 10

22 Ratio and proportion

You will learn how to:
- Use knowledge of equivalence to simplify and compare ratios (same units).
- Understand how ratios are used to compare quantities and to divide an amount into a given ratio with two parts.
- Understand and use the unitary method to solve problems involving ratio and direct proportion in a range of contexts.

Starting point

Do you remember…

- how to solve simple problems involving proportion?

  For example, a model of a car is 10 times smaller than the real car. If the model is 45 cm long, how long is the real car?

- how to use equivalent ratios?

  For example, which of these ratios is equivalent to 2 : 3?

  4 : 5   8 : 10   10 : 15

This will also be helpful when…

- you learn to simplify and compare ratios where the numbers have different units
- you work with ratios that have more than two parts
- you learn and use the relationship between ratio and direct proportion.

22.0 Getting started

The table shows the ratio of hours of daylight to hours of darkness on 21st June in different locations.

Location	Ratio of hours of daylight : hours of darkness
Buenos Aires, Argentina	7 : 17
Beijing, China	5 : 3
Sydney, Australia	5 : 7
Stockholm, Sweden	3 : 1
Nairobi, Kenya	1 : 1

What can you learn from these ratios? Write as many deductions as you can, and explain them.

Technology question Use the internet to find out the longest day and the longest night of the year in your location. What are the dates? For each of these two dates, write the ratio of hours of daylight to hours of darkness.

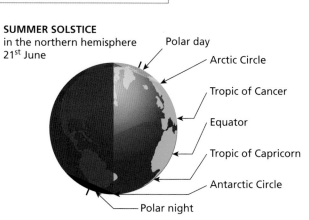

SUMMER SOLSTICE
in the northern hemisphere
21st June

Polar day
Arctic Circle
Tropic of Cancer
Equator
Tropic of Capricorn
Antarctic Circle
Polar night

22.1 Simplifying and comparing ratios

Key terms

Equivalent ratios are ratios that show the same relationship.

For example, the ratios 1 : 3 and 2 : 6 of oranges to lemons are equivalent ratios. 2 : 6 shows that there are 2 oranges for every 6 lemons, which is equivalent to 1 orange for every 3 lemons.

The **simplest form** of a ratio is the equivalent ratio that has the smallest possible whole numbers.

For example, the simplest form of 4 : 6 is 2 : 3. The simplest form of $\frac{1}{2}$: 10 is 1 : 20.

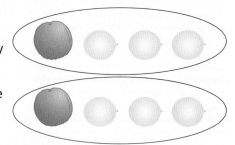

Worked example 1

Simplify each ratio.

a) 6 : 4   **b)** 18 cm : 6 cm   **c)** $1\frac{1}{2}$: 3

a) 6 : 4 = 3 : 2	6 and 4 have common factor 2. Divide both sides of the ratio by 2.	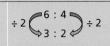	
b) 18 cm : 6 cm = 18 : 6 = 3 : 1	As both parts of the ratio are in the same unit, we can ignore the unit. To simplify in one step, divide both sides by 6. To simplify in two steps, divide both sides by 2 and then by 3.	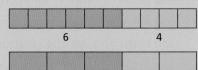	
c) $1\frac{1}{2}$: 3 = 3 : 6 = 1 : 2	Multiply the ratio by 2, so that both sides are integers. Then divide by the common factor, 3.		

Worked example 2

A paint mix is made from white and blue paints in the ratio 4 : 3.

How much blue paint is needed to mix with 120 ml of white paint?

4 : 3 = 120 ml : 90 ml 90 ml of blue paint is needed to mix with 120 ml of white paint.	Multiply both sides of the ratio to make the left side equal 120. 4 × 30 = 120, so multiply both sides by 30.	(diagram showing: 4 split into 30 30 30 30 = 120 ml; 3 split into 30 30 30 = ?) × 30 ml 4 : 3 × 30 ml 120 ml : 90 ml

Worked example 3

A paint called 'Summer green' is made by mixing blue paint and yellow paint in the ratio 6 : 5

A paint called 'Woodland green' is made by mixing blue paint and yellow paint in the ratio 4 : 3

Which shade of green is lighter?

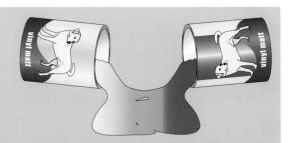

Summer green 6 : 5 = 12 : 10 Woodland green 4 : 3 = 12 : 9	Rewrite the ratios so that both have the same number on one side. One way is to multiply 6 : 5 by 2 and multiply 4 : 3 by 3. Then compare 12 : 10 and 12 : 9.
Summer green is lighter.	For the same amount of blue, Summer green has more yellow than Woodland green. (For example, for every 12 litres of blue paint, Summer green has 10 litres of yellow paint while Woodland green has only 9 litres of yellow paint.)

Exercise 1

1 Simplify the following ratios where possible.

   a) 6 : 4   **b)** 3 : 12   **c)** 30 : 40   **d)** 14 : 21

   e) 15 : 20   **f)** 40 : 16   **g)** 9 : 24   **h)** 7 : 4

2 Make three whole-number ratios that are equivalent to 8 : 12

3 Melanie looks at this diagram of coloured dots.

She makes these statements:

   a) The ratio of red to green is 1 : 1

   b) The ratio of yellow to red is 3 : 1

   c) The ratio of red to not red is 1 : 4

Is each statement true or false? Explain how you know.

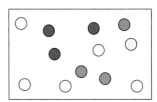

4 A farmer grows 750 kg of potatoes and 150 kg of carrots.

    a) Write down the ratio of potato mass to carrot mass in its simplest form.

    b) Write down the ratio of carrot mass to potato mass in its simplest form.

5 Here is a list of ingredients for a cake.

    a) Write down the ratio of flour to butter in its simplest form.

    b) Write down the ratio of eggs to sugar in its simplest form.

    c) Write down the ratio of sugar to flour in its simplest form.

| 240 g flour |
| 200 g butter |
| 180 g sugar |
| 150 g eggs |

6 Simplify the following ratios.

    a) $8 : 0.5$ **b)** $2.5 : 3$ **c)** $0.25 : 1$ **d)** $\frac{1}{2} : 10\frac{1}{2}$

7 Jess wants to simplify the ratio $\frac{1}{2} : 6$. She says, 'Half of 6 is 3, so the simplest form is $1 : 3$.' Is she correct? Explain your answer.

8 A drink contains 1.5 litres of orange juice and 0.5 litres of grapefruit juice. Write down the ratio of orange juice to grapefruit juice in its simplest form.

9 a) 'Steel grey' paint is made by mixing black and white paint in the ratio $5 : 6$

    'Iron grey' paint is made by mixing black and white paint in the ratio $5 : 4$

    Which paint is darker? Explain your answer.

    b) Solomon has 15 litres of black paint, and he wants to use it all to make Steel grey.

    Work out how much white paint he needs.

    c) 'Rose pink' paint is made by mixing 12 litres of red paint with 18 litres of white paint.

    'Blossom pink' paint is made by mixing 8 litres of red paint with 16 litres of white paint.

    Which paint is darker? Show your working and explain your answer.

10 a) The ratio of blue marbles to black marbles is $1 : 3$. There are 10 blue marbles. How many black marbles are there?

    b) There are 40 students on a school trip. The ratio of teachers to students is $1 : 5$. How many teachers are there?

    c) The ratio of tables to chairs in a classroom is $3 : 8$. There are 9 tables in the classroom. How many chairs are there?

> **Tip**
>
> Rewrite the ratios so that they have the same number on one side. (The ratios may not be in their simplest form.)

11 For each pair of offers, use ratios to show which is the better value.

    a) A 6 kg bag of rice for \$14, or a 15 kg bag of rice for \$30

    b) A pack of 16 pens for \$4.20, or a pack of 24 pens for \$6.00

Think about

Technology question If you type a fraction into a scientific calculator, the calculator can simplify it. How can you use this to simplify a ratio?

Discuss

If you add the same number to both sides of a ratio, why is the resulting ratio not equivalent?

Thinking and working mathematically activity

For each pair of numbers, change the first number into the second number by multiplying and/or dividing by integers. (Adding and subtracting are not allowed.)

Use the smallest possible number of steps each time.

4	→	12
15	→	3
4	→	22
20	→	8
17	→	1
9	→	23

Group the pairs by the number of steps needed. How many groups are there? In each group, what do the pairs have in common?

For some pairs, there is more than one way to change the first number to the second using the minimum number of steps. Which pairs are these? What do they have in common?

22.2 Dividing a quantity into a ratio

Worked example 4

Divide $120 in the ratio 3 : 2

	You need to split 120 into two amounts with ratio 3 : 2	120 (3 parts / 2 parts)
$3 + 2 = 5$ $\$120 \div 5 = \$24$	The ratio 3 : 2 has 5 equal parts. Divide $120 into 5 parts. Each part has value $24	120 split into 5 parts: 1 part (24), 1 part (24), 1 part (24), 1 part (24), 1 part (24)

$3 \times \$24 = \$72$ $2 \times \$24 = \$48$	3 parts are worth $3 \times \$24 = \$72$ 2 parts are worth $2 \times \$24 = \$48$	120				
		1 part	1 part	1 part	1 part	1 part
		24	24	24	24	24
The answer is $\$72 : \$48$	The two parts are $72 and $48	120				
		1 part	1 part	1 part	1 part	1 part
		72			48	

Exercise 2

1–7

1 Divide 60 in these ratios:

a) $1 : 2$ b) $11 : 1$ c) $2 : 3$ d) $1 : 3$ e) $7 : 13$ f) $3 : 7$

2 Divide 105 in these ratios:

a) $1 : 4$ b) $6 : 1$ c) $1 : 2$ d) $3 : 2$ e) $4 : 3$ f) $20 : 1$

3 A ribbon of length 65 cm is divided into two pieces in the ratio $1 : 4$. How long is each piece?

4 Diarmaid and Ingrid divide 49 berries in the ratio $3 : 4$

a) How many berries do they each get?

b) What fraction of the berries does Diarmaid get?

c) What fraction of the berries does Ingrid get?

5 Lily and Ezra share an inheritance of $25 000 in the ratio $3 : 1$

a) How much money do they each inherit?

b) What fraction of the money does Lily get?

6 A fruit drink is made by mixing juice and water in the ratio $1 : 6$. Susie wants to make 280 ml of the drink. How much juice and how much water does she need?

7 Divide:

a) $24 in the ratio $2 : 1$ b) 270 ml in the ratio $7 : 2$

c) $144 in the ratio $3 : 5$ d) 56 kg in the ratio $5 : 2$

e) $340 in the ratio $11 : 9$ f) 3 hours in the ratio $1 : 9$

> **Tip**
>
> In part **f)**, give your answer in minutes.

8 Michael and Mo want to divide $12 870 in the ratio $3 : 2$

Michael says, 'I should get $4290 and Mo should get $6435 because $12\,870 \div 3 = 4290$ and $12\,870 \div 2 = 6435$.'

Do you agree with Michael? Explain your answer.

Thinking and working mathematically activity

A box contains only circles, triangles, squares and hexagons.

- The ratio of hexagons to squares is 1 : 4
- There are 3 hexagons.
- The fraction of all the shapes that are squares is $\frac{2}{5}$
- The ratio of circles to triangles is 2 : 1

Work out how many of each shape are in the box.

Write a solution to convince someone else that your answer is correct.

Think about

How can you check your answer to a question about dividing in a ratio?

22.3 The unitary method

Key terms

The **unitary method** is a method of writing an equivalent ratio by first making one side equal 1.

For example, to write 4 : 20 in the form 3 : n, you could write 4 : 20 = 1 : 5 = 3 : 15

Worked example 5

Mikaela makes bread rolls. A batch of 8 rolls uses 320 g of flour.

How much flour would she need to make 11 rolls?

| 8 rolls : 320 g
1 roll : 40 g
11 rolls : 440 g | Write the ratio of rolls to grams of flour.

Divide the ratio by 8, to find the amount of flour for 1 roll.

Multiply the ratio by 11. | $\div 8 \left(\begin{array}{l} \text{8 rolls : 320 g} \\ \text{1 roll : 40 g} \\ \text{11 rolls : 440 g} \end{array} \right) \div 8$ $\times 11 \qquad\qquad \times 11$ |

	320 g							
1 roll	1 roll	1 roll	1 roll	1 roll	1 roll	1 roll	1 roll
40 g	40 g	40 g	40 g	40 g	40 g	40 g	40 g

	440 g									
40 g	40 g	40 g	40 g	40 g	40 g	40 g	40 g	40 g	40 g	40 g

Worked example 6

A sign at a currency exchange bureau says, 'Buy 550 Japanese yen for 5 US dollars.'

a) Amal exchanges 64 dollars for yen. Find how many yen she gets.

b) Yumi exchanges some dollars for yen. She gets 13 200 yen. Find how many dollars she exchanged.

| **a)** 550 yen : $5
110 yen : $1
7040 yen : $64 | Divide the ratio of yen to dollars by 5, to find the number of yen for 1 dollar.

Multiply the ratio by 64. | $\div 5 \left(\begin{array}{l} \text{550 yen : \$5} \\ \text{110 yen : \$1} \\ \text{7040 yen : \$64} \end{array} \right) \div 5$ $\times 64 \qquad\qquad \times 64$ |

b) 550 yen : $5	Divide the ratio of yen to dollars by 550, to find the number of dollars for 1 yen. (Keep the number of dollars in your calculator.)	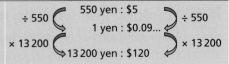
1 yen : $0.09...		
13 200 yen : $120	Multiply by 13 200.	

Exercise 3

1 The cost of 5 notebooks is 350 cents.

a) Find the cost of 1 notebook.   b) Find the cost of 7 notebooks.

2 The mass of 6 AA batteries is 138 g.

Which statement below is not correct? Correct this statement.

a) The mass of 15 AA batteries is 345 g.

b) The mass of 5 AA batteries is 115 g.

c) The mass of 3 AA batteries is 69 g.

d) The mass of 19 AA batteries is 414 g.

> **Tip**
>
> Start by finding the mass of one battery.

3 a) You need 240 g of flour to make 12 cupcakes. Calculate the amount of flour needed to make 15 cupcakes.

b) Chris buys 7 pencils for 56 cents. Calculate the cost of 6 pencils.

c) A cleaner charges $72 for 6 hours of cleaning. How much does the cleaner charge for 9 hours of cleaning?

d) 20 miles is approximately the same distance as 32 km. Use this fact to work out the approximate number of miles in 8 km.

4 Joe is paid $55 for 10 hours' work at a supermarket. To calculate how much he gets paid for 2 hours, he does the calculations below.

$55 for 10 hours

$110 for 20 hours

$11 for 2 hours

a) Is Joe correct?

b) Show how to do the calculation by first finding the pay for 1 hour.

c) Use two different methods to calculate how much Joe will be paid for 3 hours of work.

5 The height of 6 identical bricks is 72 cm.

Which method of finding the height of 18 of these bricks is not correct? Correct it.

a) Find the height of 1 brick, then multiply the result by 18.

b) Multiply 72 cm by 3.

c) Find the height of 2 bricks, then multiply the answer by 9.

d) Find the height of 3 bricks, then multiply the answer by 3.

6 Khaled exchanged 160 euros for dollars. He received $180.

   a) Makoto exchanged 400 euros for dollars. How many dollars did he receive?

   b) Chae-rin exchanged some euros and received $135. How many euros did she pay?

7 80 paperclips have a mass of 160 g.

Amy and Tom buy 360 g of paperclips. To find out the number of paperclips they bought, they do different calculations but reach the same answer. Are both calculations correct?

Amy's answer	Tom's answer
1 paperclips = 80 ÷ 160 = $\frac{1}{2}$ g	1 paperclip = 160 g ÷ 80 = 2 g
$\frac{1}{2}$ × 360 = 180 paperclips	360 g ÷ 2 g = 180 paperclips

> **Tip**
>
> Use a bar model to explain your answer.

Thinking and working mathematically activity

The unitary method can be used to solve all of the problems below – but it is not always the most efficient method.

For which of the problems is the unitary method the most efficient method? Explain your choices. (You do not have to solve the problems.)

A. The cost of 7 identical items is $42. What is the cost of 3 of these items?

B. The cost of 8 identical items is $104. What is the cost of 16 of these items?

C. The cost of 6 identical items is $72. What is the cost of 2 of these items?

D. The cost of 15 identical items is $90. What is the cost of 25 of these items?

E. The cost of 2 identical items is £52. What is the cost of 3 of these items?

F. The cost of 12 identical items is $264. What is the cost of 18 of these items?

For the problems where the unitary method is not the most efficient, write solutions using a more efficient method.

Compare your work with other students' work, and discuss any differences.

Consolidation exercise

1 Write each ratio in its simplest form.

   a) 55 : 40   **b)** 80 : 32   **c)** 7 : 0.5   **d)** $1\frac{1}{2}$: 5   **e)** 0.25 : 1.25   **f)** 0.5 : $\frac{1}{4}$

2 Which is the odd one out? Explain your answer.

A 32 : 48	B 4 : 6	C 108 : 162	D 56 : 72	E 10 : 15

3 Graham is dividing $90 between two people in the ratio 2 : 3

Graham says, 'The first person will get $\frac{2}{3}$ of the money.'

Do you agree with Graham? Explain your answer.

4 The total mass of 20 tiles is 12 kg. Calculate:

a) the mass of 2 tiles   b) the mass of 7 tiles   c) the mass of 1000 tiles.

5 A sign at a currency exchange bureau says: 'Buy 180 Chinese yuan for 25 Swiss francs'.

a) Sven exchanges 300 Swiss francs for Chinese yuan. How many yuan does he receive?

b) Bridget exchanges some Swiss francs for Chinese yuan. She receives 900 yuan. How many francs did she exchange?

End of chapter reflection

You should know that...	You should be able to...	Such as...
You can simplify a ratio by dividing the numbers in the ratio by the same factor. You can write a whole-number ratio in its simplest form by dividing the numbers in the ratio by the highest common factor.	Simplify whole-number ratios.	Write each ratio in its simplest form: a) $6 : 21$ b) $35 : 14$
The simplest form of a ratio cannot include a fraction or decimal.	Simplify ratios that include fractions or decimals.	Write each ratio in its simplest form: a) $0.5 : 9$ b) $5 : 3\frac{1}{4}$
When you share a quantity in a given ratio, you must divide the quantity by the total number of parts.	Divide a quantity into two parts in a given ratio.	Share $45 in the ratio $7 : 8$
The unitary method means finding the value of one part or one unit first.	Use the unitary method to solve simple problems involving ratio and proportion.	A shop is selling 5 tins of beans for $3.00. How much would 12 tins of beans cost?

23 Probability 2

You will learn how to:

- Design and conduct chance experiments or simulations, using small and large numbers of trials. Analyse the frequency of outcomes to calculate experimental probabilities.

Starting point

Do you remember…

- how to find the theoretical probability of a particular outcome happening?

  For example, a bag contains three red buttons, two green buttons and five orange buttons. A button is chosen at random. What is the probability that the button is red?

- how to make a tally chart?

  For example, what frequency does ||||| || represent?

This will also be helpful when…

- you learn more about recording outcomes of events to estimate probabilities.

23.0 Getting started

This is a game for two players

You will need:

- Two six-sided dice
- A grid like this to record the number of wins for each player

Player	Number of wins
Player A	||||| |||||
Player B	|||||

Rules:

Players take turns to roll the two dice and add together the scores.

- Player A wins if the total score is 6, 7, 8 or 9. Otherwise Player B wins.
- Play the game 15 times and record how many times each person won.

As a class, discuss which player won more frequently.

Combine the results of everyone in the class – which player seems to have a higher chance of winning?

23.1 Probability from experiments

Key terms

A **chance experiment** is an activity where many outcomes are possible, like rolling a dice. An experiment is designed that repeats an event or a series of events many times. We record the outcomes of all the events and use these data to estimate the probability.

We use an **estimate of probability** when it is not possible to use theoretical probability. This estimate of probability is called **relative frequency.**

For example, you would use relative frequency as an estimate if you wanted to find the probability of a getting head on a biased coin.

The relative frequency of a particular outcome

$$= \frac{\text{number of times that outcome occurred}}{\text{number of trials}}$$

The more times an experiment is repeated, the closer the relative frequency comes to an estimate of the **theoretical probability**.

> **Did you know?**
>
> Probability theory is used in sports such as football to analyse the strengths and weaknesses of teams.

Worked example

Ruth has a four-sided dice with sides numbered 1, 2, 3 and 4. She has rolled the dice 20 times and wants to use the results to estimate the probability of rolling a 4.

Number on the dice	1	2	3	4
Frequency	7	8	2	3

a) Find the experimental probability that Ruth will roll a 4 using this dice.

b) Write down how Ruth could get a better estimate of the probability.

a) Experimental probability $= \dfrac{\text{number of 4's rolled}}{\text{total number of rolls}}$ $= \dfrac{3}{7 + 8 + 2 + 3}$ $= \dfrac{3}{20}$	Calculate the relative frequency to find an estimate of the experimental probability. Divide the number of 4s rolled by the total number of rolls. To find the total number of rolls, add all of the frequencies. The estimated probability is 0.15 as a decimal and as a percentage it is 15%.	
b) Ruth could get a better estimate of the probability by rolling the dice more times.	Increasing the number of times an experiment is repeated usually gives estimated probabilities that are closer to the theoretical values.	

Exercise 1

1 Lakmi records the number of wins, draws and losses for his basketball team in one season.

Win	Draw	Lose
36	25	19

a) How many matches were played in total?

b) Based on these results, find the relative frequency of Lakmi's team winning their next match.

2 Mrs Lopez travels to work by train. This is a record of how often the train she catches is on time, is early and is late.

On time	Early	Late
43	4	3

Based on these results, estimate the probability that:

a) the train will be on time

b) the train will be early

c) the train will be late.

3 Minoo has a restaurant. Customers can choose to have their meals with rice, with potato or with neither. These were the customer choices one day:

Rice	Potato	Neither
29	13	1

Based on these choices, find the relative frequency of a customer choosing a meal:

a) with rice

b) with potato

c) with neither rice nor potato.

4 Traffic Watch is a company that surveys road traffic.

They record the traffic passing their camera between 9 a.m. and 5 p.m.

These are the results for Monday:

Car	Bicycle	Lorry
19	8	8

Based on these results, find an estimate of the probability that:

a) the vehicle passing the camera will be a lorry

b) the vehicle passing the camera will be a bicycle

c) the vehicle passing the camera will be a car.

5 A biased six-sided dice, with sides numbered 1, 2, 3, 4, 5 and 6 respectively, is thrown 200 times.

The results are shown in the table below.

a) Copy and complete the table.

Number on dice	1	2	3	4	5	6
Frequency	48	50	48	22	22	10
Relative frequency	0.24			0.11		

b) What is an estimate of the probability of the dice showing a 3?

c) What is an estimate of the probability of the dice showing an odd number?

d) What is an estimate of the probability of the dice showing a prime number?

6 Felix makes a spinner like the one shown. He spins it 70 times.
The table shows his results.

Outcome	Yellow	Green	Blue
Frequency	8	19	43

a) Use the results to find the relative frequency of the spinner landing on green.

b) Explain how Felix could get a more accurate estimate of the probability.

7 Lydia and Natasha perform an experiment to investigate whether a four-sided dice is biased. The dice is numbered 1 to 4.

The table shows the data collected.

	Total rolls	1	2	3	4
Lydia	50	14	18	2	16
Natasha	50	11	23	5	11

Lydia and Natasha are surprised their results are different.

a) Explain why getting different results when an experiment is repeated is not unusual.

b) Combine both sets of results to calculate the relative frequencies of each number.

c) Is the dice biased? Explain your answer.

d) How could Lydia and Natasha be more certain of their answer to part **c)**?

8 For each situation below, write down whether the theoretical probability can be found OR the probability should be estimated from an experiment.

a) The probability of winning a game of Go against a computer.

b) The probability of rolling an even number on a fair dice.

c) The probability that the next car you see will be black.

d) The probability of picking a green counter from a bag containing 4 red counters and 8 green counters.

> **Discuss**
>
> If the weather forecast says that there is a 70% chance of rain tomorrow, is this an estimated experimental probability or a theoretical probability?

9 A spinner with three different colour sections is spun 50 times.
It lands on pink 12 times, on purple 23 times and on black 15 times.

   a) Giving your answers as decimals, find the experimental probability of the spinner landing on:

   i) a pink section   **ii)** a purple section   **iii)** a black section.

   b) Which colour section do you think is the largest?

10 A teacher puts some black counters and some red counters into a bag. Sadia and Hala are asked to find the experimental probability that a counter taken from the bag at random is black.

   Sadia takes out a counter at random, notes the colour and then replaces the counter into the bag.

   She does this 50 times and gets 32 black counters.

   a) What is Sadia's relative frequency of choosing a black counter?

   Hala then carries out the same process but repeats it 200 times. She gets 104 black counters.

   b) What is Hala's relative frequency of choosing a black counter?

   c) Which of the two relative frequencies is likely to be closer to the theoretical probability? Explain your answer.

11 A fair spinner has six sections, each a different colour.

   Jana spins the spinner 100 times and it lands on the red section 18 times.

   Sara spins the same spinner 100 times. Will she also find that the spinner lands on red exactly 18 times? Give a reason for your answer.

12 A bank tries to answer all telephone calls within 1 minute.

   Mädchen tests this by phoning her bank at different times of day.

   Mädchen results are shown in the table.

	Morning	Afternoon	Evening
Number of calls made	76	90	63
Number of times answered within 1 minute	50	53	45

   At what time of day is the bank most likely to answer a call within 1 minute? Show your working.

Thinking and working mathematically activity

Stick a small amount of sticky tack to the tails side of a coin. Work with a partner to design an experiment to estimate the probability of getting a tail when your coin is tossed.

Before you start, discuss with your partner:

- how many times you should toss the coin
- how you should present your results
- if it matters what type of coin is used.

After the experiment is complete, discuss with your partner:

- how you could improve the accuracy of your results
- what you think would happen if you changed the amount of sticky tack on the coin
- whether you would get the same results if you repeated the experiment.

Consolidation exercise

1 **a)** Chris rolls a dice 50 times. These are her results:

Number on dice	1	2	3	4	5	6
Frequency	8	9	9	7	9	8
Estimated probability						

    i) Copy and complete Chris's table showing the estimated probability for each number. Give your answers as fractions in simplified form.

    ii) From your answers, do you think Chris is using a fair dice?

b) Lesley rolls another dice 50 times. These are her results:

Number on dice	1	2	3	4	5	6
Frequency	9	14	8	8	3	8
Estimated probability						

    i) Copy and complete Lesley's table showing the estimated probability for each number. Give your answers as fractions in simplified form.

    ii) From your answers, do you think Lesley is using a fair dice?

c) If Lesley and Chris are playing a game with a dice, whose dice should they use?

2 A drawing pin is thrown into the air and lands either on its point or with the point facing up.

Mora is in Class 8. Her class of 30 students including Mora each throw a drawing pin 200 times and record their individual results. Mora's drawing pin lands point up 120 times.

a) What is the relative frequency of Mora's drawing pin landing point up?

The class then add their results. Out of 6000 throws, they found that the drawing pin lands point up 3012 times.

> **Tip**
>
> To divide by 6000, divide by 6 first and then by 1000.

b) From the whole class results, what is the relative frequency of their drawing pins landing point up?

c) Which of these two estimates of probability is likely to be more accurate? Give a reason for your answer.

3 All the students at a school either walk, cycle, get the school bus or travel in a car to get to school.

The table shows how a sample of boys and girls, chosen at random, travel to school most often.

	Walk	Cycle	Bus	Car	Total
Boys	25	35	25	15	
Relative frequency				0.15	
Girls	52	56	54	38	
Relative frequency	0.26				

a) Copy and complete the table.

b) State whether the following statements are true or false based on the results in the table and give a reason for your decision.

   i) A girl, chosen at random from the sample, is more likely than a boy to come to school by car.

   ii) A girl, chosen at random from the sample, is more likely than a boy to cycle to school.

   iii) A girl, chosen at random from the sample, is more likely than a boy to take the bus to school.

   iv) Girls and boys are equally likely to walk to school.

4 Jake and Peter each spin the same spinner 20 times. The results are shown in the table below.

	Lands on red	Lands on blue	Lands on green
Jake	6 times	6 times	8 times
Peter	7 times	8 times	5 times
Combined total			

a) For each of the following statements say whether it is true, may be true or is false. If it is false then correct the statement.

   i) The estimated experimental probability of Jake's spinner landing on blue is 0.3

   ii) The estimated experimental probability of Peter's spinner landing on blue is 0.4

   iii) If you combine the two sets of results then the estimated experimental probability of the spinner landing on blue is 0.7

   iv) The spinner is equally likely to land on any of the three colours.

   v) There are only three colours on the spinner.

b) If the experiment was repeated, would you expect to get the same results? Give a reason for your answer.

End of chapter reflection

You should know that...	You should be able to...	Such as...
Relative frequency can be used to estimate the probability of an event from a series of trials. Relative frequency of a particular outcome $= \frac{\text{number of times that outcome occurred}}{\text{number of trials}}$ If an experiment or trial is repeated different results can happen. A larger number of trials will give better results.	Estimate an experimental probability by calculating the relative frequency.	Use the results in the table to find the experimental probability of the spinner landing on each colour. **Colour**: red, blue, green **Spins**: 12, 16, 12 How could more accurate experimental probabilities be obtained?

24 Sequences

You will learn how to:
- Understand term-to-term rules and generate sequences from numerical and spatial patterns (linear and integers).
- Understand and describe nth term rules algebraically (in the form $n \pm a$, $a \times n$ where a is a whole number).

Starting point

Do you remember…

- that an integer is a whole number, positive or negative?

  For example, 12 and –5 are integers, but 20.5 is not.

- how to count on in whole numbers from different starting points?

  For example, counting on in 6s from 4 gives 4, 10, 16, …

- looking at patterns in number squares?

  For example, looking at the pattern made by multiples of 5 in a 10 × 10 number square.

This will also be helpful when…

- you learn about straight line graphs
- you learn about more complicated sequences and number patterns
- you learn how to generalise results.

24.0 Getting started

Katie is investigating multiplication by 5.

She takes a 2-digit number, adds the digits and multiplies the answer by 5.

She writes the answer then repeats the process with the new number.

She continues to do this until she finds a pattern.

The first number she tries is 15. She gets 15, 30, 15, 30, …

Then she tries 24 and gets these results:

24, 30, 15, 30, …

What do you notice about both patterns?

What do 15 and 24 have in common?

Which other numbers give a pattern which goes back to 30 and 15?

Try other numbers.

Which starting number gives the most numbers in the pattern before you find one that repeats?

Katie's method	
Start :	15
$1 + 5 = 6$	
$6 \times 5 = 30$	30
$3 + 0 = 3$	
$3 \times 5 = 15$	15
$1 + 5 = 6$	
$6 \times 5 = 30$	30
…	…

24.1 Generating sequences

Key terms

A **sequence** is a set of numbers, shapes, letters or objects placed in an order that makes a pattern or follows a rule.

Each item in a sequence is called a **term**.

In the sequence 1, 3, 5, 7, 9, 11, ... the first term is 1 and the second term is 3. The dots after the 11 mean that the sequence continues using the same rule.

The **position number** of a term tells you where it is in the sequence. The first term has position number 1, and the sixth term has position number 6. In the sequence 1, 3, 5, 7, 9, 11, ... the number 9 has position number 5.

When you make a sequence by doing something to each term to get the next term, you are using a **term-to-term rule**.

For example, in the sequence 1, 3, 5, 7, 9, 11, ... you can get from one term to the next by using the rule 'add 2'.

Worked example 1

The first term of a sequence is 5 and the term-to-term rule is 'add 6'.
Find the first six terms of this sequence.

5, 11, 17, 23, 29 and 35	The term-to-term rule tells you to start with 5 and add 6 to get the next term, $5 + 6 = 11$ Add 6 again. $11 + 6 = 17$	5 ↘+6 11 ↗+6 17 ↗+6 23 ↗+6 29 ↗+6 35
	Continue to get the first 6 terms: $17 + 6 = 23$ $23 + 6 = 29$ $29 + 6 = 35$	

Exercise 1

1 The table gives the first term and the term-to-term rule of a sequence.

Find the first five terms of each sequence.

	First term	Term-to-term rule
a)	2	Add 3
b)	5	Add 2
c)	−10	Add 5
d)	50	Subtract 4
e)	4	Subtract 4
f)	−12	Subtract 3

2 Find the next two terms in each of these sequences.

a) 4, 13, 22, 31, …   b) 52, 45, 38, 31, …   c) −4, −9, −14, −19, …

3 Write the term-to-term rule for each sequence and use it to find the 8th term.

a) 5, 8, 11, 14, …   b) 10, 7, 4, 1, …

c) −120, −110, −100, −90, …   d) −15, −20, −25, −30, …

4 Find the missing terms in these sequences.

a) 5, 9, ___ , ___ , 21, 25, …   b) 56, 49, ___ , 35, ___ , 21, …   c) ___ , −7, −10, ___ , −16, …

5 In this sequence, the first three terms are correct, but one of the other terms is incorrect and does not follow the term-to-term rule:

2, 5, 8, 11, 14, 18, 20, 23, …

a) What is the term-to-term rule?

b) What is the value of the third term?

c) Write the number that does not follow the term-to-term rule.

d) What number should it be?

e) Find the tenth term.

6 Maria and Emil both think of a sequence.

Maria's sequence
−5, −2, 1, 4, …

Emil's sequence
Term-to-term rule = 'subtract 4'

The eighth term of both sequences is the same. Find the first four terms of Emil's sequence.

7 A sequence begins 42, 48, 54, 60, …

Julian says that 99 is in the sequence.

Explain how you know that Julian is not correct.

Thinking and working mathematically activity

Choose a number between −10 and 10.

Make up your own sequences that have your chosen number as the second term.

Write down the term-to-term rule for each sequence.

Make sure you include at least two sequences for each type of term-to-term rule:

| Add …. | | Subtract … |

Worked example 2

Claire is making triangle patterns using 10 cm sticks to make equilateral triangles. Each time she adds another triangle, she writes down how many sticks she uses and the perimeter of her new shape.

She records her results in a table.

Position number	1	2	3	4	5
Number of sticks	3	5			
Perimeter (cm)	30	40			

a) Complete the table.

b) Write the term-to-term rule for the number of sticks used to make the pattern.

c) Write the term-to-term rule for the sequence of numbers for the perimeter.

d) How many sticks will be needed to make the 7th pattern in the sequence?

e) What will the perimeter be for the 7th pattern?

---

a)

Position number	1	2	3	4	5
Number of sticks	3	5	7	9	11
Perimeter (cm)	30	40	50	60	70

Count the total number of sticks to get the middle row of the table.

Count the sticks around the outside of each shape and multiply by 10 (because each stick is 10 cm long) to get the bottom row of the table.

Draw the fifth pattern and count the sticks.

---

b) add 2

The number of sticks increases by 2 each time, so the term-to-term rule must be add 2.

---

c) add 10 (cm)

The perimeter increases by one stick each time, and one stick is 10 cm, so the term-to-term rule for the perimeter must be add 10.

d) 15 sticks	For the 5th pattern you need 11 sticks, so using the rule: 6th pattern = 11 + 2 = 13 7th pattern = 13 + 2 = 15	
e) 90 cm	For the 5th pattern the perimeter is 90 cm, so using the rule: 6th pattern = 70 + 10 = 80 cm 7th pattern = 80 + 10 = 90 cm	

Exercise 2

1 These L-shaped patterns are made by adding one square to the top and one to the right side of the L each time.

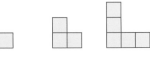

Pattern 1　　Pattern 2　　Pattern 3

This table records the number of squares used in each pattern.

Position number	1	2	3	4	5
Squares					

a) Copy and complete the table for Patterns 1 to 5.

b) Write the term-to-term rule for the sequence you have generated.

c) Find the number of squares in Pattern 8.

2 The first four patterns in a sequence are shown below.

a) Draw the fifth pattern, then copy and complete the table.

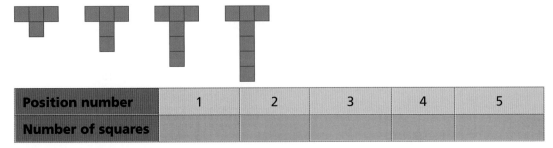

Position number	1	2	3	4	5
Number of squares					

b) How many squares are there in the 10th pattern?

c) How many squares are in pattern 100? Explain how you worked out your answer.

3 These are the first three patterns in a sequence.

   a) Draw the next two patterns in the sequence.

   b) Create a table showing the position number and the number of squares in the pattern for the first five shapes.

   c) Find a term-to-term rule for this sequence.

   d) How many squares are there in the 20th pattern?

   e) How many squares are there in the 100th pattern? Give a reason for your answer.

   f) Describe how you could find the number of squares in any pattern in the sequence.

4 These are the first four patterns in a sequence.

   a) Draw the fifth pattern, then copy and complete the table.

Position number	1	2	3	4	5
Number of squares	4				

   b) What is the term-to-term rule for the number of squares?

   c) Find the number of squares in the 10th pattern.

   d) Amy says the number of squares in the 20th pattern must be twice the number in the 10th pattern. Is she correct? Explain your answer.

5 These are the first three patterns in a sequence. Each pattern is made up of rectangles and squares.

   a) Draw the fourth and fifth patterns, then copy and complete the table.

Position number	1	2	3	4	5
Number of squares	1	2			
Number of rectangles	2	3			

   b) What is the term-to-term rule for the number of rectangles?

   c) What is the term-to-term rule for the number of quadrilaterals in each pattern? Use the term-to-term rules for the number of squares and rectangles to explain why your rule works.

6 These patterns represent arrangements of people sitting at tables. The squares represent the tables and a person sits at each available side. Tables are added in straight lines, as shown.

Pattern 1   Pattern 2   Pattern 3

a) Draw Pattern 4.

b) Copy and complete the table showing the number patterns.

c) Find a term-to-term rule for the number of people in each pattern.

Position number	1	2	3	4
Number of tables				
Number of people				

d) How many people and tables would there be in the 20th pattern? Explain how you found your answer.

Thinking and working mathematically activity

Decide on a term-to-term rule and work out the first eight terms of your sequence.

Now cross out every other term starting with the first term.

What is the term-to-term rule for your new sequence?

Would the rule be the same if you crossed out every other term starting with the second term?

Investigate with other term-to-term rules. What do you notice?

Did you know?

The Fibonacci sequence is a famous number pattern where you add the two previous terms to get the next one:

1, 1, 2, 3, 5, 8, 13, 21, 34, 55, …

$1 + 1 = 2$

$1 + 2 = 3$

$2 + 3 = 5$

$3 + 5 = 8$

and so on …

Fibonacci wrote about the sequence in 1202, but mathematicians knew about the sequence before then.

Fibonacci numbers can be found in nature and art, and are also used in computing and economics.

Can you think of other examples where you might find the Fibonacci sequence in nature?

24.3 Position-to-term rules

Key terms

The **general term** of a sequence is a way of describing how you can find the value of a term from its position number.

A **position-to-term** rule is the rule which relates the position number to the value of the term.

Did you know?

The position number of a term tells you where it is in the sequence. The first term has position number 1, the second has position number 2 and the sixth term has position number 6.

For example, look at this sequence.

Position number	1	2	3	4	5	6
Term	5	6	7	8	9	10

The position-to-term rule is to add 4 to the position number to get the value of the term.

The 200th term would be 200 + 4 = 204

A position-to-term rule can also be written using algebra.

For the example above, for a position number n, the general term would be $n + 4$. This way of writing a position-to-term rule is called an **nth term** rule.

Worked example 3

Reema is making patterns with toothpicks.

Look at the patterns she has made.

a) How many toothpicks will Reema need for the 5th and 6th patterns?

b) Copy and complete the table to show the number of toothpicks in each diagram.

Pattern number (position)	1	2	3	4	5	6
Number of toothpicks						

c) Describe how the number of toothpicks in each pattern is connected to the position of the pattern in Reema's sequence.

d) Write the rule for the nth term of Reema's sequence.

a) 5th pattern: 15 toothpicks

6th pattern: 18 toothpicks

Look at the patterns you already have and count the toothpicks:

1st : 3 toothpicks

2nd : 6 toothpicks

3rd: 9 toothpicks

4th: 12 toothpicks

The term-to-term rule is 'add 3'.

5th pattern:

6th pattern:

b)

Pattern number (position)	1	2	3	4	5	6
Number of toothpicks	3	6	9	12	15	18

Put the number of toothpicks in each pattern into the table.

c) Multiply the pattern number by 3 to get the number of toothpicks.	$1 \times 3 = 3$ $2 \times 3 = 6$ $3 \times 3 = 9$ $4 \times 3 = 12$ $5 \times 3 = 15$ $6 \times 3 = 18$	

Pattern number (position)	1×3	2×3	3×3	4×3	5×3	6×3
Number of toothpicks	3	6	9	12	15	18

d) nth term = $3n$	To get the term in the sequence you multiply the position number by 3, so when the position number is n, the term will be $3 \times n$, which is $3n$.	$1 \times 3 = 3$ $2 \times 3 = 6$ $3 \times 3 = 9$ $n \times 3 = 3n$

Exercise 3

1 Copy and complete the table for each of these position-to-term sequence rules:

Position	1	2	3	4	5	6
Term						

a) Each term in the sequence is four more than its position number.

b) The position-to-term rule is to multiply by 5.

c) Each term in the sequence is 5 less than its position number.

d) The position-to-term rule is to multiply by 10.

2 A sequence begins 7, 14, 21, 28, … . Describe how you could find any term from its position number.

3 Each term in a sequence is six more than its position number. Find the 8th term of the sequence.

4 The position-to-term rule of a sequence is multiply by 9. Find the first four terms of the sequence.

5 Match each position-to-term statement with its algebraic expression for the nth term.

Multiply the position number by 4 to find the term value.	$n + 3$
Find the term by adding 3 to the position number.	$n - 4$
Subtract 4 from the position number to find the term value.	$6n$
The term value is 6 times the position number.	$4n$
Find the term value by subtracting 6 from the position number.	$n - 6$

6 The nth term of a sequence is $n - 2$. Find the 7th term of this sequence.

7 The nth term of a sequence is $12n$. Find the first four terms of this sequence.

8 The nth term of a sequence is $n + 11$. Find the sum of the first term and the sixth term.

9 Find the nth term rule for each of these sequences.

a) 6, 7, 8, 9, 10, …

b) 4, 8, 12, 16, 20, …

c) −2, −1, 0, 1, 2, …

d) 20, 40, 60, 80, …

e) 11, 12, 13, 14, …

f) 15, 30, 45, 60, …

10 Eli is making patterns with square tiles.

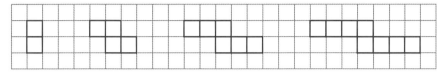

a) Find the nth term rule for his sequence.

b) How many tiles will he need for the 10th pattern in the sequence?

c) Eli has 75 tiles altogether. Will he have enough tiles to be able to make the 40th pattern in his sequence? Explain how you know.

Thinking and working mathematically activity

Here are three cards that all show the same sequence.

Number pattern	term-to-term rule	nth term rule
2, 4, 6, 8, …	add 2	$2n$

This is a three-way snap.

Make four more sets of cards that show a three-way snap.

Join other students to make a group of three or four. Put all your cards together and shuffle them.

Deal the cards so that each person has six cards.

Put the rest of the cards face down and turn the top card over.

Take it in turns: You can either swap the face-up card for one of your cards, or take a card from the pack which you can then either swap for one of your cards or put on the face-up pile.

You win when you have two complete three-way snaps.

Consolidation exercise

1 Find the next two terms in these sequences.

a) 4, 7, 10, 13, …

b) 6, 12, 18, 24, 30, …

c) 4, 1, −2, −5, …

d) 20, 40, 60, 80, …

2 Write the term-to-term rule for these sequences and find the 10th term for each sequence.

a) −12, −5, 2, 9, …

b) 500, 450, 400, 350, …

3 Misha makes a sequence, but one of the numbers is wrong.

4, 11, 18, 24, 32, 39, 46, …

a) Which number is wrong? Explain how you know.

b) Write the sequence correctly and write the term-to-term rule.

c) Misha thinks that 81 is in his sequence. Is he correct? Explain how you know.

d) Misha now makes another sequence. Find the term-to-term rule for his new sequence.

6, 13, 20, 27, 34, 41, 48, ….

e) Compare the two sequences. Write one thing that is similar and one thing that is different about them.

4 These are the first three patterns in a sequence.

a) Draw the fourth and fifth patterns, then copy and complete the table.

Position number	1	2	3	4	5
Number of squares	8				

b) What is the term-to-term rule for the number of squares?

c) Find the number of squares in the 15th pattern.

d) Can there ever be an odd number of squares in a pattern in this sequence? Explain how you know.

5 Each term in a sequence is 10 more than its position number.
What is the 15th term in this sequence?

6 The position-to-term rule is to multiply by 6. Find the first five terms of the sequence.

7 Match each sequence with the correct nth term rule:

5, 10, 15, 20, …	$n - 12$
−4, −3, −2, −1, ….	$15n$
−11, −10, −9, −8, ….	$n + 12$
15, 30, 45, 60, …	$5n$
13, 14, 15, 16, …	$n + 7$
8, 9, 10, 11, 12, …	$n - 5$

8 Alex is making flower arrangements for a celebration.

There are 7 flowers in each arrangement.

Each table will have 3 arrangements.

a) Copy and complete the table.

Number of tables	1	2	3	4	5
Number of arrangements	3	6			
Number of flowers	21				

b) How many arrangements will be needed for six tables?

c) How many flowers will Alex need for six tables?

d) What is the nth term rule for the number of arrangements?

e) What is the nth term rule for the number of flowers?

f) Alex has 200 flowers in total. Does she have enough flowers to do all the arrangements for ten tables? Explain how you know.

End of chapter reflection

You should know that...	You should be able to...	Such as...
A term-to-term rule is a rule telling you how to find the next term in a sequence from the previous term.	Find the next term in a sequence if you know the term-to-term rule.	The third term in a sequence is 23. If the term-to-term rule is 'subtract 5', what are the fourth and fifth terms of the sequence?
Sequences can be generated from a spatial pattern. A spatial pattern is an arrangement of objects.	Find terms in a sequence from a spatial pattern and be able to describe the term-to-term rule.	These are the first four patterns in a sequence. **a)** Copy and complete this table showing the number of squares in each pattern. | Position	1	2	3	4	5
---	---	---	---	---	---
Squares	1	3	5			 **b)** Write down the term-to-term rule for this sequence. **c)** Find the 10th term of this sequence.
A position-to-term rule is a relationship between the position of a term in a sequence and the value of the term.	Find terms from a position-to-term rule.	For a sequence whose position-to-term rule is 'multiply by 3' **a)** Find the 20th term of the sequence. **b)** Write the algebraic rule for the nth term of the sequence.

25 Accurate drawing

You will learn how to:
- Draw parallel and perpendicular lines, and quadrilaterals.
- Use knowledge of scaling to interpret maps and plans.
- Visualise and represent front, side and top view of 3D shapes.

Starting point

Do you remember...

- how to use a ruler and draw lines to the nearest centimetre and millimetre?

  For example, draw a line that is 4.7 cm long.
- how to use a protractor to measure and draw an angle?

  For example, use your protractor to draw a 45 degree angle.
- how to recognise parallel and perpendicular lines?

  For example, which of these lines is parallel to line A and which is perpendicular to line C?

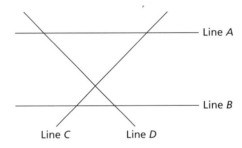

- how to use ratio notation?

  For example, an orange paint is made by mixing red and yellow paint in the ratio 1 : 4.

  Sheila use 2 litres of red paint. How much yellow paint will she need to make the orange paint?
- the properties of common 3D shapes?

  For example, a triangular prism has two triangular faces and three faces that are rectangles.
- that 3D objects like cuboids, prisms and pyramids can be represented by nets and can be drawn on isometric paper?

  For example, can you draw a net for this cuboid?

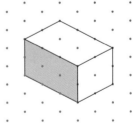

This will also be helpful when...

- you learn to solve problems about bearings
- you use maps in real-life
- you find the surface area of 3D shapes.

25.0 Getting started

The diagram shows an arrangement of four cubes. It has been drawn on isometric paper.

- Make different arrangements using four cubes. Record each of your arrangements on isometric paper. Try to ensure you have found all the possible different arrangements.
- Find all the different arrangements that can be made using five cubes. Record your arrangements on isometric paper.

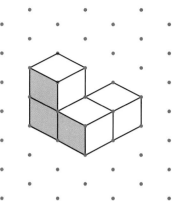

25.1 Construction of parallel and perpendicular lines

Key terms

A **set square** is an tool used to provide a straight edge at a right angle to a line.

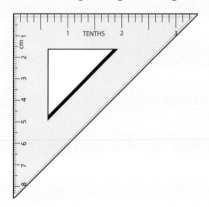

Worked example 1

Use a ruler and set square to construct:

a) a pair of perpendicular lines

b) a pair of parallel lines 4 cm apart

a) Begin by drawing a base line.

Put your set square on the base line as shown. The set square makes a 90° angle.

> **Tip**
>
> Your 90° angle does not have to be at the end of the base line – you could draw a perpendicular line part way along the base line.

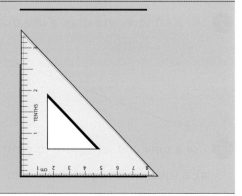

b) Begin by drawing a line. This will be one of your parallel lines.

Put your set square on the line you have drawn.

Place your ruler against the set square with the 0 mark on the line you have drawn.

Hold your ruler still and slide the set square up the ruler to the 4 cm mark.

Draw along the base of the set square.

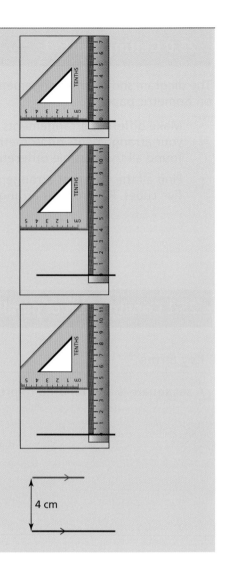

4 cm

> **Did you know?**
>
> Set squares are not only used to find 90° angles.
>
> Set squares are used by architects and engineers to create technical drawings.

Exercise 1

1 Using a ruler and a set square, construct:

    a) a pair of perpendicular lines

    b) a pair of parallel lines, 6 cm apart

    c) a pair of parallel lines, 3.5 cm apart.

> **Think about**
>
> Do lines that are perpendicular have to touch/cross?

2 Use a set square to draw a line that is perpendicular to AB and passes through point C.

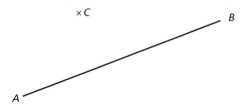

3 Use a ruler and set square to construct:

    a) a square with side length 7 cm

    b) a square with side length 4.5 cm

4 Use a ruler and set square to construct these rectangles.

a)
3 cm
7 cm

b)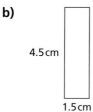
4.5 cm
1.5 cm

c)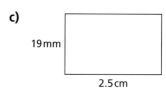
19 mm
2.5 cm

5 Use a ruler, set square and protractor to construct these quadrilaterals.

a)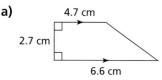
4.7 cm
2.7 cm
6.6 cm

b)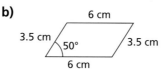
6 cm
3.5 cm
50°
3.5 cm
6 cm

Think about

Technology question
Use mathematical drawing software to try drawing the quadrilaterals in question 5.

c)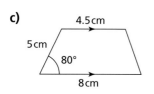
4.5 cm
5 cm
80°
8 cm

d)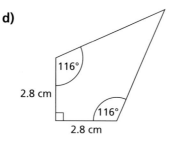
116°
2.8 cm
116°
2.8 cm

6 Here is a sketch of a trapezium.

Kudzai says that the perimeter of the trapezium is 25 cm.

Use a ruler, protractor and set square to construct this trapezium accurately. Comment on Kudzai's claim.

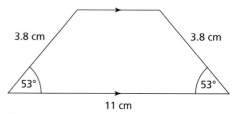
3.8 cm
3.8 cm
53°
53°
11 cm

Thinking and working mathematically activity

Construct this isosceles trapezium and this parallelogram using a ruler and protractor.

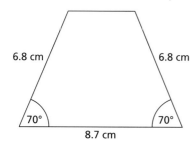
6.8 cm
6.8 cm
70°
70°
8.7 cm

5 cm
5 cm
53°
127°
7 cm

Show that it is possible to draw a circle around the outside of the isosceles trapezium so that it passes through all four vertices.

Is it possible to draw a similar circle around the outside of the parallelogram?

Try this with other isosceles trapeziums and parallelograms. Make some conclusions.

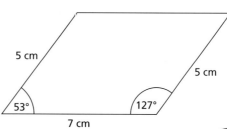

25.2 Scale drawings

A map or object that is drawn **to scale** has each dimension in proportion to the original.

Drawings of very large objects are a **reduction** of their actual size.

This car is drawn so that 1 cm represents $\frac{1}{2}$ metre in real-life.

The scale is 1 cm : $\frac{1}{2}$ m

We can write this without units as 1 : 50

Drawings of very small objects are an **enlargement** of their actual size.

Sometimes we draw things bigger than real-life. Here is a drawing of an ant.

The drawing of the ant is 5 times bigger than in real-life. The scale is 5 : 1

Key terms

To scale means in proportion so that each length on the drawing or model is in proportion with the same length in real-life.

A **map scale** tells us the relationship or ratio between a length on a map or drawing, and the length in real-life. For example, if 1 cm on a map represents 5 km, the scale is 1 : 500 000

A **plan** is a drawing or sketch of an object or place as it would be seen from above, also called a birds-eye view.

Worked example 2

a) This drawing of a door has a scale of 1 cm : 40 cm. What is the real height and width of the door? Give your answers in metres.

b) The scale of a map is 1 : 25 000. A road measures 875 m.

What is the length of this road on the map?

c) A scale drawing is made of a car. The car is 4.5 m long in real-life. The car is 9 cm long on the drawing.

Find the scale used to make the scale drawing.

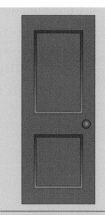

a) 1 cm : 40 cm
= 1 cm : 0.4 m

The width of the door is 2 cm on the drawing. The real width will be
2 × 0.4 = 0.8 m

The height of the door is 5 cm on the drawing. The real height will be
5 × 0.4 = 2 m

The scale is 1 : 40, meaning that each centimetre on the drawing represents 40 cm = 0.4 m in real-life.

Measure the length and width of the door on the drawing.

To get the real lengths in metres, **multiply** the measured lengths by 0.4.

Width:

1 cm	1 cm
0.4 m	0.4 m

Height:

1 cm	1 cm	1 cm	1 cm	1 cm
0.4 m	0.4 m	0.4 m	0.4 m	0.4 m

| | b) 1 cm : 25 000 cm
= 1 cm : 250 m
875 ÷ 250 = 3.5
The length of the road on the map is 3.5 cm. | The scale 1 : 25 000 means that each centimetre on the map represents 25 000 cm in real-life.

25 000 cm = 250 m, so the scale is 1 cm : 250 m.

To get the length of the road on the map, **divide** the length on the map by 250. |

1 cm	1 cm	1 cm	$\frac{1}{2}$ cm
250 m	250 m	250 m	125 m

c) 9 cm represents 4.5 m
= 450 cm

The scale is 9 : 450

Divide by 9 to simplify this to 1 : 50

Write both measurements in the same unit (here centimetres).

Form a ratio using these measurements.

Write the ratio in its simplest form.

	9 cm							
1	1	1	1	1	1	1	1	1

	450 cm							
50	50	50	50	50	50	50	50	50

Exercise 2

1 Jack makes a scale drawing of a building. In real-life the building is 7 m high, and in the drawing it is 14 cm high. Work out the scale.

2 On a scale drawing, one centimetre represents eight centimetres in real-life.

Calculate the real-life distance represented by:

a) 7 cm   **b)** 4.5 cm   **c)** 25 cm   **d)** 4 cm

3 The floor plan of a school library is drawn with a scale of 1 cm : 40 cm

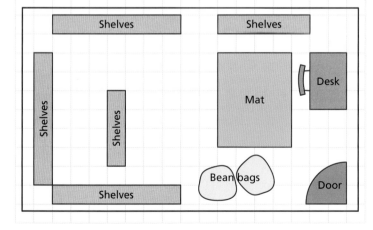

> **Tip**
>
> Don't forget to leave space for chairs!

a) Find the length and width of the desk.

b) Find the length and width of the mat.

c) How many metres of the walls are covered by shelves?

d) The library wants four new computer workstations. Each workstation is 80 cm by 120 cm. Redesign the library to fit the new computer workstations.

4 A scale model is made of a park using a scale of 1 cm to represent 10 m. The table shows a list of measurements on the scale model and the equivalent real measurements of parts of the park.

Copy and complete the table.

		Scale model	Real park
a)	Length of park		500 m
b)	Width of park	35 cm	
c)	Width of pond		30 m
d)	Height of oak tree	0.5 cm	

5 Write the scale for these maps. Give your answers without units.

a) 1 cm on the map is equal to 0.5 m in real-life.

b) 2 cm on the map is equal to 2 m in real-life.

c) 5 cm on the map is equal to 25 m in real-life.

6 Victor has four plan drawings. The scales are:

1 : 250　　　　1 : 10 000　　　　1 : 1000　　　　1 : 100 million

Match each scale to the map it fits best.

a) world map　　**b)** map of a town　　**c)** map of a school　　**d)** a design for a model volcano

Explain your answers.

7 Match the scale to the real-life measurement.

1 : 10　　　　　　1 cm represents 100 m

1 : 100　　　　　1 cm represents 10 m

1 : 1000　　　　1 cm represents 1 m

1 : 10 000　　　1 cm represents 10 cm

8 Johanna draws a scale drawing of a sports club. The scale is 1 : 500

Find the real-life length of:

a) the tennis court, which is 4.8 cm on the scale drawing. Give your answer in metres.

b) the shower, which is 1.7 mm on the scale drawing. Give your answer in centimetres.

9 Tim finds a crab on the beach that is 18 cm wide and 15 cm high. He draws a scale drawing of it using a scale of 1 : 3. What is the height and width of the crab in his drawing?

10 Filipe is 1.6 m tall. This drawing shows him standing next to his house.

a) Find the scale of the drawing.

b) Find the height of his house.

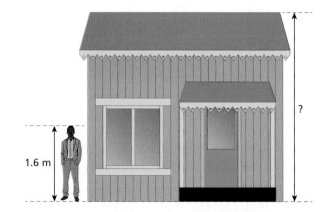

1.6 m

11 A map has a scale of 1 : 10 000

    a) What actual distance does 1 cm on the map represent? Give your answer in metres.

    b) The length of a road on the map is 5 cm. How long is the road in metres?

    c) The distance between two villages on the map is 25 cm. What is the actual distance between the villages in km?

    d) The length of the road from the church to the top of the hill is 800 m. What length, in cm, on the map represents the road?

12 The diagram (not drawn to scale) shows a park which is 500 metres long and 600 metres wide. It has a triangular play area in one corner which has one side of 200 m and another side of 300 m, as shown in the diagram. At the other end of the park there is a small rectangular car park 200 m long and 100 m wide.

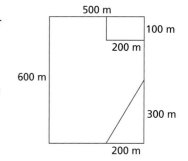

    a) Make a scale drawing of the park and its features using a scale of 2 cm to represent 100 m.

    b) What is the shortest distance between the play area and the car park on the scale drawing?

13 A map has a scale of 1 : 200 000

    a) A park on the map has length 1.6 cm. Find the length of the park in real-life.

    b) The width of a forest is 8.8 km. Find the width of the forest on the map.

14 Mae has a map with a scale of 1 : 500 000

The distance between two villages on the map is 3 cm. Mae thinks this means that the villages are 1.5 km apart in real-life.

Is Mae correct? Give a reason for your answer.

Thinking and working mathematically activity

Dax wants to put out a fire.

He needs to get from his current position to the river to fill up his bucket.

He then will go directly to the fire.

- One route is shown on the scale drawing. The scale is 1 : 5000. Find the distance he would run if he used this route.

- Show that this is not the shortest route that Dax could use.

- What is the shortest route you can find?

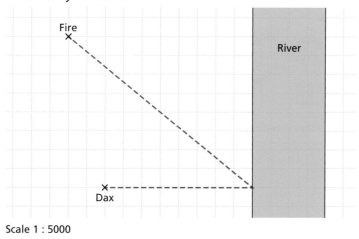

Scale 1 : 5000

25.3 Plans and elevations

Key terms

The **plan view** is the view of a 3D shape seen from the top.

An **elevation** is the view of a 3D shape seen from the front or the side.

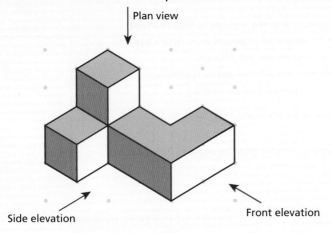

Plan view

Side elevation   Front elevation

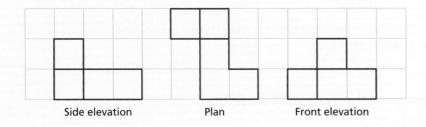

Side elevation   Plan   Front elevation

Worked example 3

Here is a prism.

a) Draw the front and side elevations of this prism.

b) Draw the plan view of the prism.

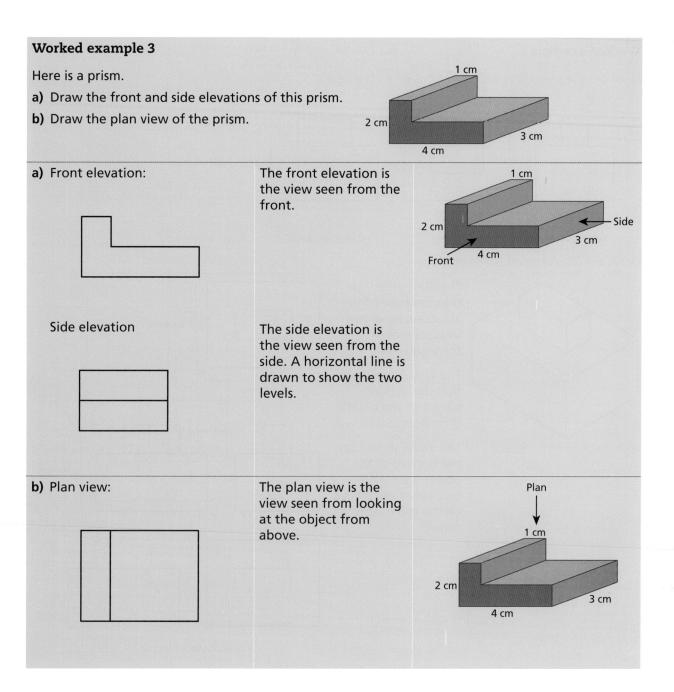

a) Front elevation:

The front elevation is the view seen from the front.

Side elevation

The side elevation is the view seen from the side. A horizontal line is drawn to show the two levels.

b) Plan view:

The plan view is the view seen from looking at the object from above.

Worked example 4

Here are the plan view and elevations of a 3D shape.

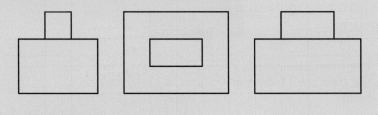

Side elevation   Plan view   Front elevation

Draw the 3D shape on isometric paper.

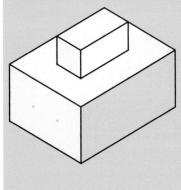

The plan shows us that the base of the 3D shape measures four squares by three squares.

The elevations show that the overall height of the object is three squares.

The views show that the bottom of the object is a cuboid. Two cubes are placed on the top of this cuboid.

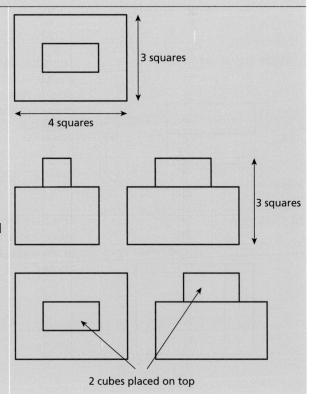

3 squares

4 squares

3 squares

2 cubes placed on top

Exercise 3

1 Draw the front and side elevations of these objects.

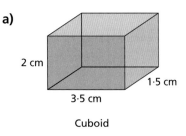

a)

2 cm
3·5 cm
1·5 cm

Cuboid

b)

4 cm
3 cm
6 cm

Prism with isosceles triangle as cross-section

c)

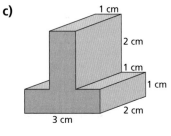

1 cm
2 cm
1 cm
1 cm
3 cm
2 cm

Prism with T-shaped cross-section

2 Draw the plan view of these objects.

a)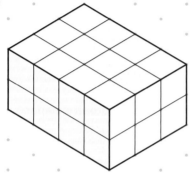

2·5 cm

Cube

b)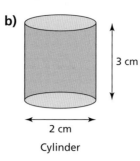

3 cm

2 cm

Cylinder

c)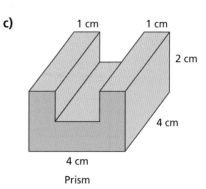

1 cm   1 cm

2 cm

4 cm

4 cm

Prism

3 These objects are made from centimetre cubes. Draw the front elevation, side elevation and plan view of each object. Make each of your views full size.

a)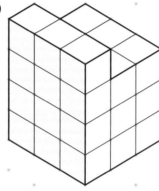

b)

c)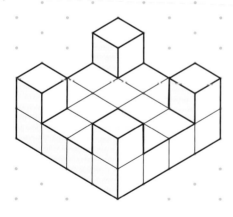

d)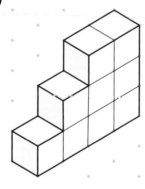

4 Draw the plan view, front elevation and side elevation of each of these 3D objects. Draw each view full size.

a)

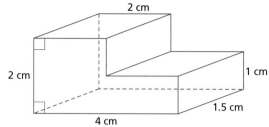

b)

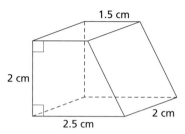

c)

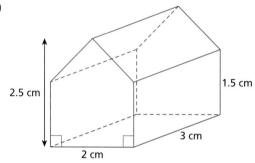

d)

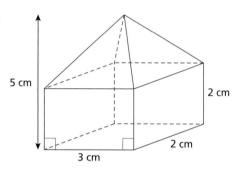

5 Here are the plans and elevations of some objects. Draw each object on isometric paper.

a)

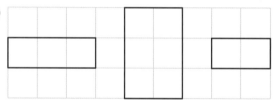

Side elevation   Plan   Front elevation

b)

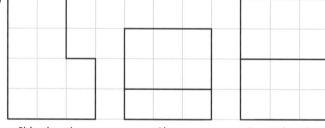

Side elevation   Plan   Front elevation

c)

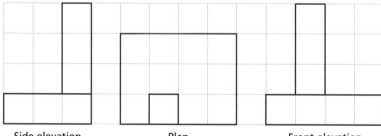

Side elevation   Plan   Front elevation

6 Here are the side elevation and plan view of an object made from cubes.

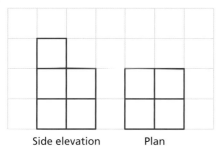

Side elevation   Plan

Tip

In question 6, try using cubes to build objects that have the given views. Can you make one that is made from fewer than ten cubes?

Alex says that the object must contain 10 cubes. Show that Alex is not correct..

Thinking and working mathematically activity

These diagrams show the front elevation, side elevation and plan of three objects.

a)
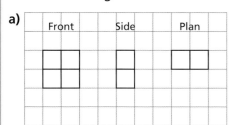

b)
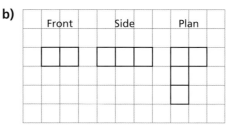

c)
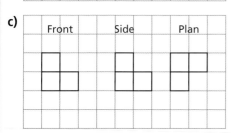

Make each of the objects from cubes and draw them on isometric paper.

Consolidation exercise

1 Use a ruler, protractor and set square to draw these shapes accurately.

a)
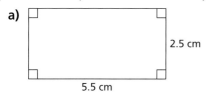
2.5 cm

5.5 cm

b)
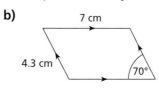
7 cm

4.3 cm

70°

2 Which would be a good scale for a drawing of your classroom: 1 : 10, 1 : 100 or 1 : 1000? Give reasons for your answer.

3 A scale drawing is made using a scale of 2 cm to 5 m. Write this scale as a ratio in its simplest form.

4 Laz reads a map with a scale of 1 : 40 000. On the map he sees there is a town 12 cm away from where he is. He thinks this shows the town to be about 5 km away. Is Laz correct? Explain your answer.

5 For each of these diagrams, draw **(i)** the plan **(ii)** the side elevation and **(iii)** the front elevation.

a)
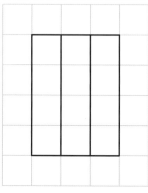

b)
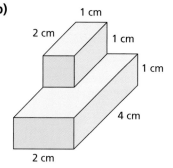

c)
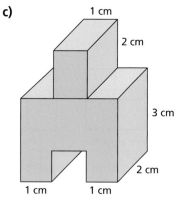

6 Here is the front elevation of a 3D object.

Front elevation

Draw on isometric paper two possible representations of the object.

End of chapter reflection

You should know that...	You should be able to...	Such as...
A set square can be used to draw parallel and perpendicular lines.	Draw parallel and perpendicular lines.	Draw one line parallel to AB and one line perpendicular to AB. A ———— B
A scale is expressed as a ratio.	Interpret map scales. Make a scale drawing from information that you have been given and understand scales written in different ways.	What does 1 : 10 000 mean? A map has a scale of 1 : 10 000. The distance between the bank and the garage is 2 cm on the map. What is the real distance in metres?
An object can be represented by drawing a plan view as well as side and front elevations.	Draw the plan and elevations for a 3D object.	Draw the plan, side elevation and front elevation for this 3D object.

26 Thinking statistically

You will learn how to:
- Record, organise and represent categorical, discrete and continuous data. Choose and explain which representation to use in a given situation:
  - Venn and Carroll diagrams
  - tally charts, frequency tables and two-way tables
  - dual and compound bar charts
  - waffle diagrams and pie charts
  - frequency diagrams for continuous data
  - line graphs
  - scatter graphs.
- Use knowledge of mode, median, mean and range to describe and summarise large data sets. Choose and explain which one is the most appropriate for the context.
- Interpret data, identifying patterns, within and between data sets, to answer statistical questions. Discuss conclusions, considering the sources of variation, including sampling, and check predictions.

Starting point

Do you remember...

- the different types of data?

  For example, name something about you that is categorical, something that is continuous and something that is discrete.

- how to draw different types of graph to represent data?

  For example, draw a scatter graph to see if there is any connection between your test scores in two subjects.

- how to find the mean, median, mode and range for a set of data?

  For example, find the mean and range of these test scores.

36%	48%	64%	91%	72%	38%
55%	25%	56%	83%	62%	47%

This will also be helpful when...

- you decide which table you use to record your data
- you decide which graph or chart you use to display your data.

26.0 Getting started

Irene's garden is full of slugs.

She wants to get rid of them to allow her vegetables to grow.

She collects the slugs each day and takes them to a friend who researches slugs and their behaviour.

She records the number of slugs she collects each day and enters the total in a spreadsheet each week.

Here are her results.

	A	B	C	D	E
1	Date	08-Apr	15-Apr	22-Apr	29-Apr
2	Number of slugs collected	225	148	314	222
3					
4	Date	06-May	13-May	20-May	27-May
5	Number of slugs collected	178	87	161	92
6					
7	Date	03-Jun	10-Jun	17-Jun	24-Jun
8	Number of slugs collected	140	137	176	107
9					
10	Date	08-Jul	15-Jul	22-Jul	29-Jul
11	Number of slugs collected	116	120	54	21
12					
13	Date	05-Aug	12-Aug	19-Aug	26-Aug
14	Number of slugs collected	45	111	63	25
15					
16	Date	02-Sep	09-Sep	16-Sep	23-Sep
17	Number of slugs collected	76	26	44	23
18					

Irene wants to know if what she is doing has any effect on the number of slugs in her garden.

What statistical diagram would you draw to help her? What calculations might you do?

Draw a diagram and do some calculations to help you offer advice to Irene.

26.1 Choosing appropriate graphs and tables

Worked example 1

Here are the marks for a History test for 30 students.

24 31 31 35 19 37

29 32 25 34 36 17

32 29 30 28 37 38

39 34 27 30 33 26

28 20 23 22 31 31

a) Create a grouped frequency table for the marks.

b) Draw an appropriate diagram to show the information in your table.

c) Comment on the distribution of marks.

a)

Mark	Tally	Frequency
15 – 19	II	2
20 – 24	IIII	4
25 – 29	NN II	7
30 – 34	NN NN I	11
35 – 39	NN I	6

The lowest mark is 17 and the highest is 39, so grouping the marks based on just the tens digit would give just three groups.

That is not enough groups to show how the distribution varies.

Grouping in 5s gives the five groups shown.

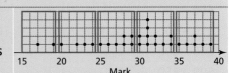

b)

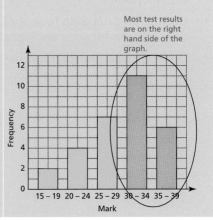

A bar chart is chosen so the frequencies can be easily seen and compared.

A pie chart would lose that data and it is harder to compare areas of sectors than heights of bars.

The data are discrete so the bars should not touch.

c) The bar chart shows the marks are distributed unevenly, with more students scoring at least 30 than scoring under 30.

Perhaps the students found it a fairly easy test!

Look at the shape of the graph. The bars are higher on the right-hand side of the graph.

Most test results are on the right hand side of the graph.

Chapter 26: Thinking statistically   **303**

Thinking and working mathematically activity

Emi recorded the mass of plastic she collected in her street during April.
Here are the masses recorded to the nearest gram.

479	434	164	132	574	460
386	360	506	670	604	151
339	473	586	226	284	233
626	686	209	458	379	503
460	260	599	468	173	231

She grouped the data as follows.

Mass (g)	Tally	Frequency
100 – 150	I	1
150 – 200	III	3
200 – 300	Ң I	6
300 – 400	IIII	4
400 – 450	I	1
450 – 500	Ң I	6
500 – 600	Ң	5
600 – 650	II	2
650 – 700	II	2

> **Did you know?**
>
> It takes plastic bottles up to 450 years to biodegrade.
>
> Plastic bags can take from 10 years to 1000 years to decompose in landfill.
>
> To try to reduce the use of plastics, some companies now use bags made of potato starch instead. These will decompose in a normal compost bin.

She drew the following bar chart.

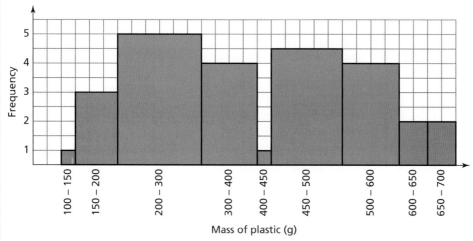

- Write anything that is wrong or could be misleading about the grouped frequency table and the diagram.

- Use the original data to write a better grouped frequency table.
  Give reasons for your decisions.

- Use your new grouped frequency table to draw an appropriate diagram.
  Give reasons for your decision.

Exercise 1

 1 Petra is conducting a survey of the number of people who attend an art exhibition. She records the number of people each day for 24 days.

Here are her results:

75	54	89	61	59	75	74	81	86	79	90	59
60	66	68	71	70	65	73	75	82	80	63	72

a) Construct a grouped frequency table.

b) Draw an appropriate diagram to show the information in your table.

c) Comment on the distribution of people at the art exhibition.

 2 Kim is marking essays written by her students. The number of words in the essays written by her students is shown below.

495	511	483	502	500	496	532	498	496
499	503	521	487	518	526	508	514	503

a) Construct a grouped frequency table to show this data.

You should aim to have about six class intervals in your table.

b) Explain these chosen intervals.

c) Draw an appropriate diagram to show the information in your table.

d) Explain the chosen diagram.

e) How many words do you think Kim asked the students to write? Give a reason for your answer.

3 Kira records the time it takes her kettle to boil when it contains different amounts of water.

Amount of water (litres)	0.5	0.7	0.8	0.9	1	1.2	1.5
Time (seconds)	124	170	185	212	235	290	348

a) Draw an appropriate diagram to show these data.

b) Comment on the relationship between the amount of water in the kettle and the time it takes to boil.

 4 Julia surveyed the students in her class to find out how many bicycles their families own.

Here is her data.

Females: 3 4 4 6 2 5 4 3 5 3 4 3 2 4 4

Males: 2 4 6 2 2 2 3 5 4 4 4 2 2 6 6

a) Draw an appropriate diagram to show these data.

b) Write a reason why you chose your diagram.

5 Veer did a pocket money survey of 40 students in his youth club.

Here is his data. Each amount was given to the nearest dollar.

Boys	30, 32, 25, 14, 26, 26, 37, 33, 39, 18, 36, 18, 34, 20, 25, 10, 10, 16, 25, 13
Girls	44, 46, 30, 16, 44, 42, 32, 41, 40, 31, 20, 45, 33, 27, 21, 40, 34, 27, 34, 45

a) Construct a grouped frequency table for the data.

b) Draw an appropriate statistical diagram for the data.

c) Write a reason why you chose your diagram.

d) Veer states that the boys get less pocket money than the girls.
   Write a reason why the diagram supports Veer's statement.

6 The table shows the temperature of a cup of hot water at different times.

Time since boiled (minutes)	0	5	10	15	20	25
Temperature (°C)	100	79	61	47	37	30

a) Draw a statistical diagram for the data.

b) Write a reason why you chose your diagram.

c) Write down what your diagram shows you about how the temperature of the water changes.

d) Find how long it will take for the temperature to drop to 40 °C.

26.2 Choosing an appropriate average

Some averages are better than others in different situations. The mean is often considered the best average because it is calculated using all the data. However, the median is a more appropriate measure of average when the data contains one or more unusually large or unusually small values.

> **Think about**
>
> If a data set contained a value that was much larger than the other data values, what effect would that data value have on the value of the mean? Why would it not affect the value of the median?

Worked example 2

Decide whether the mode, median, mean or range should be used to answer the following.

a) Find the most common colour of European snakes.

b) Find the average height of the members in a band.

c) Find which athlete throws the most consistent distances with a javelin.

d) Find the average salary of ten workers, nine of which are manual labourers and one is the managing director.

a) The data is non-numerical. The only average you can use for non-numerical data is the mode.	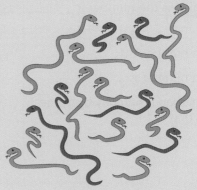 Mode = green
b) The data are numerical. The mean may be the best average to use as it takes into account the heights of all the band members.	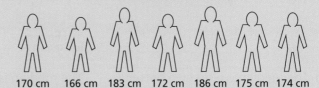 170 cm 166 cm 183 cm 172 cm 186 cm 175 cm 174 cm Mean height = 175.1 cm (1dp)
c) You are looking at how spread out the distances of each javelin throw are for each athlete. Use the range. The athlete with the smallest range is the most consistent.	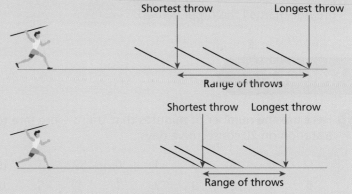
d) The median may be the best average to use as the managing director could earn a significant amount more than the manual labourers.	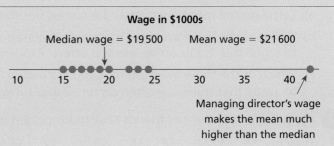

Worked example 3

The frequency table shows the number of attempts 30 people needed to throw a ball into a bucket.

Number of attempts	1	2	3	4	5	6	7	8	9	10
Frequency	15	9	4	1	0	0	0	0	0	1

a) Find the median number of attempts.

b) Explain why the median is a better measure of average than the mean in this context.

a) The median is the number in position $\frac{n+1}{2} = \frac{30+1}{2} = 15.5$

The 15th number is 1.

The 16th number is 2.

So the median is halfway between 1 and 2, that is 1.5

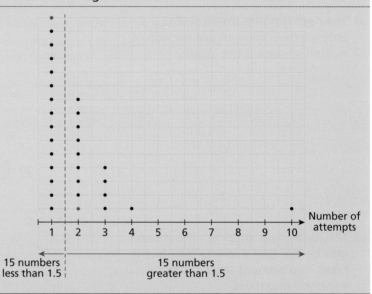

b) The median would be a better measure of average because the mean would be affected by the person who took 10 attempts (much greater than all the other values).

The median is a better measure of average than the mean if there is a data value that is much larger (or smaller) than the rest.

Exercise 2

> **Did you know?**
>
> A value in a data set that is very different to the others is called an outlier (or an anomalous value).

1 Here are the number of minutes that trains were late to arrive at a station on 20 consecutive days:

22	8	6	0	1	0	52	1	10	0
0	1	2	0	7	1	1	5	0	0

a) Calculate the mean, median and mode of this data set.

b) Which of the three averages that you calculated in part a) would be best to use if you wanted to argue that trains arrive late too often? Explain your answer.

c) Which of the three averages that you calculated in part a) would be best to use if you wanted to argue that trains are often on time? Explain your answer.

2 Aaliyah asks eight of her friends how much pocket money they receive each week.

They say: $5 $7.50 $6 $8 $3.50 $7 $4.50 $8.50

Aaliyah's mum will give her the average amount of pocket money that her friends receive.

Do you think Aaliyah will use the mean or the median average?

Give a reason for your answer and state any figures which you calculate.

3 Write down whether the mode, median or mean should be calculated in each circumstance.

   a) Rian wants to find the average test score on a maths exam.

   b) Jane wants to find out the average eye colour of her classmates.

   c) Mamadu took part in a gymnastics competition. He wants to know whether he came in the top half of all of the athletes who took part.

4 Eight swimmers take part in a sponsored swim for charity. The amounts below show how much money they each raised.

$26 $87 $52 $106.50 $23.25 $47.52 $134.80 $10.50

The local paper is going to publish the average amount of money raised by the swimmers.

Should the paper use the mean or the median average?

Give a reason for your answer and show your calculations.

5 An art gallery has 15 paintings for sale. Their prices (in $) are listed below.

230	415	95	12 300	550	815	245	85
400	309	110	415	330	265	380	

a) Calculate an appropriate average for the data.

b) Give a reason for your choice of average.

6 Julian runs a bookshop. The table shows the number of books purchased by each of his last 25 customers.

Number of books	1	2	3	4	5	6	7	8	9	10
Number of customers	8	9	4	2	1	0	0	0	0	1

   a) Calculate the median number of books purchased by these customers.

   b) Calculate the mean number of books purchased by these customers.

   c) Why is the median a better choice of average than the mean for the data?

7 Zenab runs a club. The table shows the number of people attending her club over the past 24 weeks.

Number of people	12	13	14	15	16
Number of weeks	5	7	6	4	2

   a) Calculate an appropriate average for the data.

   b) Give a reason for your choice of average.

> **Did you know?**
>
> Life expectancy in the UK is calculated using the mean. However, using the median would give a figure of about 2–3 years higher and using the mode would give a figure of about 6 years higher.

Thinking and working mathematically activity

Dom has type 1 diabetes. He tests his blood regularly to monitor the amount of insulin he needs to inject.

He checks his blood sugar level before each meal and records his results.

Here are his results for one month. The units are millimoles per litre (mmol/L).

Day	1	2	3	4	5	6	7	8	9	10
Before breakfast	6.7	11.4	7.1	8.4	9.6	5.5	7.0	12.2	9.6	7.0
Before lunch	7.4	7.1	6.8	4.1	6.9	7.4	7.1	9.8	9.6	9.7
Before dinner	11.9	3.5	17.0	10.3	5.9	10.9	11.3	8.7	3.7	9.6

Day	11	12	13	14	15	16	17	18	19	20
Before breakfast	7.2	6.3	4.7	9.8	11.9	8.1	7.2	12.3	7.5	9.3
Before lunch	9.4	8.6	11.9	5.7	11.2	10.4	9.5	9.2	6.9	8.3
Before dinner	8.9	12.2	5.6	9.3	5.4	10.9	6.7	10.6	13.0	8.6

Day	21	22	23	24	25	26	27	28	29	30
Before breakfast	6.4	9.6	4.7	10.8	10.8	10.6	10.5	6.7	7.1	5.7
Before lunch	6.6	2.7	12.8	3.6	11.7	5.5	9.3	7.4	8.2	6.3
Before dinner	5.3	9.1	11.4	9.6	8.4	7.2	4.2	8.1	5.6	7.4

The doctor would like Dom's average results to be in the range 3.9 to 7.8 mmol/L.

Dom has a hypothesis that there is no difference in his average results before breakfast, lunch or dinner.

Test his hypothesis over a suitable number of consecutive days.

If you were Dom, what conclusions could you draw?

26.3 Project

Did you know?

The word hypothesis comes from Greek and means 'to suppose'. The plural of hypothesis is hypotheses.

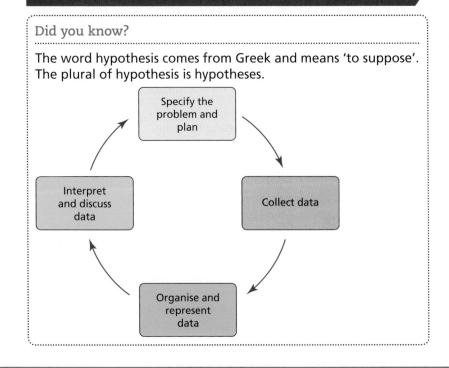

Key terms

A **hypothesis** (or **conjecture**) is an idea that you want to investigate by collecting data.

This is the **Statistical Enquiry Cycle**.

When you do a statistical project, you start at the top of the cycle by specifying the problem or hypothesis and making a plan.

Then you collect the data you need by using a questionnaire or data collection sheet.

Next, you organise the data by drawing appropriate diagrams and making appropriate calculations.

Finally, you need to interpret and discuss the data. Depending on your results you may accept or reject your hypothesis.

Sometimes your data generate new paths of enquiry. If you had problems with your hypothesis or collecting the data, you can go back the top of the cycle, modify the hypothesis and investigate further.

Thinking and working mathematically activity

How much plastic is being thrown away in your classroom? In your school? In your home?

How much of it is recycled?

How can you find out?

1. Working with a partner, discuss what issue you will focus on.
2. Write a hypothesis on what you expect to find.
3. Decide what data you will collect. Consider how you will record your data and how you will display your findings.
4. Collect an appropriate amount of data.
5. Summarise your data, giving reasons for your choice of diagrams. If you have access to technology, use it.
6. Make suitable calculations, giving reasons why you chose them.
7. Write your conclusions. How does your data compare with national and international recycling levels? Will your results have an impact on your school or on how you recycle at home?
8. If you repeated the project, what would you change?

Consolidation exercise

1 a) Draw an appropriate diagram to represent the average weekly temperature in a town over a year.

b) Write a reason why you chose your diagram.

Temperature (°C)	Number of weeks
10 – 15	2
15 – 20	6
20 – 25	10
25 – 30	15
30 – 35	19

2 Frankie has nine text books on his shelf. The table shows the thickness and mass of each book.

Thickness (mm)	28	6	12	16	20	23	10	21	8
Mass (g)	1480	265	720	1080	870	1220	535	990	335

a) Draw an appropriate diagram to show the data.

b) Comment on what your diagram shows.

c) Find an appropriate average value for the mass of the books.

d) Give a reason for your choice of average.

3 A shop sells boxes of medium eggs and large eggs. The table shows the number of boxes of each size of egg bought on Monday and Tuesday.

Size of egg	Monday	Tuesday
Medium	18	14
Large	25	23

a) The shop owner thinks about drawing a pie chart to show the information.
   Give a reason why a pie chart may not be the most appropriate choice of diagram.

b) Draw a more appropriate diagram to show the data.

4 Pat recorded the number of texts he sent each day during a month. The table shows his data.

Number of texts	0	1	2	3	4	5	6	7	8
Number of days	1	0	0	2	1	4	10	7	6

a) Calculate an appropriate average for the data.

b) Give a reason for your choice of average.

c) Give a reason why the range is not a good way to measure the spread of the data.

5 Cherri plays darts. She wants to join the local darts team.
These are 12 totals she scored with three darts:

16, 18, 84, 83, 90, 39, 29, 18, 105, 33, 76, 82.

a) Write down which average she would want to put in her team application form.
   Give a reason for your answer.

b) The team captain wants to know the range of any 11 totals, so she can see how consistent Cherri is.
   State which total Cherri should ignore. Give a reason for your answer.

End of chapter reflection

You should know that...	You should be able to...	Such as...
There are many ways that statistical data can be represented.	Decide on an appropriate diagram to show a set of data.	Here are the number of hours of sunshine each day in a month at a holiday resort: 7, 8, 11, 10, 12, 10, 8, 7, 12, 10, 8, 10, 7, 8, 11, 9, 12, 10, 10, 11, 8, 9, 10, 10 7, 7, 9, 9, 8, 10 Draw an appropriate diagram to illustrate this data. Give a reason why you chose your diagram.
Mode, median, mean and range are used to describe large data sets.	Choose and explain the most appropriate average for the content.	Find an appropriate average number of hours of sunshine per day for the above data.
The statistical enquiry cycle helps you research a project involving data.	Apply the statistical enquiry circle when doing a project involving data.	Undertake a project into people's journey to school.

27 Relationships and graphs

You will learn how to:
- Understand that a situation can be represented either in words or as a linear function in two variables (of the form $y = x + c$ or $y = mx$), and move between the two representations.
- Use knowledge of coordinate pairs to construct tables of values and plot graphs of linear functions, where y is given explicitly in terms of x ($y = x + c$ or $y = mx$).
- Recognise straight line graphs parallel to the x- or y-axis.
- Read and interpret graphs related to rates of change. Explain why they have a specific shape.

Starting point

Do you remember...

- a coordinate is a point specified by a pair of numbers (x, y)?

  For example, the point (3, 2) has a position 3 squares across from the origin and 2 squares up.

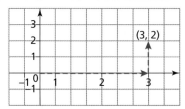

- how to plot coordinates?

  For example, plot the point (4, −1).

- how to substitute into a formula?

  For example, substitute $t = 4$ into the formula $m = 4t$.

- how to read scales?

  For example, on this travel graph, each square on the horizontal axis represents 15 minutes.

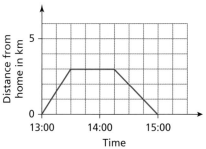

This will also be helpful when...

- you learn how to plot graphs of more complex functions
- you need to interpret graphs in science and other subjects, as well as in mathematics.

27.0 Getting started

Coordinate game

A game for 2 or more players.

You will need:
- two sets of cards numbered from −5 to 5
- a copy of the axes shown for each player.

How to play:

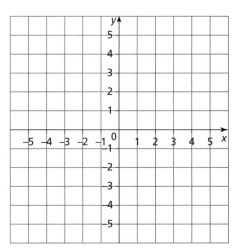

- Each player draws three lines on their axes: one horizontal, one vertical and one at an angle.
- Players take turns to choose a card at random from each set. For example, if the cards selected were 2 and 3, the coordinates could be (2, 3) or (3, 2).

- Players mark the point on their grid with a cross. Check that each player has plotted their point in the right place – if it is wrong, erase it and move it to the correct position.
- The cards are replaced, ready for the next player to have their turn.
- A player wins when they get three coordinates on any of their lines.

To increase your chance of winning, think about what lines would be good lines to use. Which lines would be less good choices?

27.1 Linear relationships

Key terms

For any function, the input and output numbers can be used to create a **coordinate pair**, (x, y).

For example, for the function $y = x + 4$, when you put the number 2 in, you get 6 out, which would give the coordinate pair (2, 6).

When you have a **linear function**, the coordinate pairs created by the function form a straight line when plotted on a coordinate grid (linear means 'straight line').

Graphs of linear functions often have practical uses, such as when you are converting from one currency to another. These are called **conversion graphs**.

Did you know?

Coordinates were invented by a French mathematician and philosopher, René Descartes, in 1637. They are sometimes called Cartesian coordinates after him. Coordinates can also be used in three dimensions to show a point in space.

Worked example 1

An inch is a unit of measurement of length. 1 inch is approximately equal to 2.5 cm.

a) Write a function connecting length in inches (x) and length in cm (y).

b) Complete the table for inches and centimetres.

Inches, x	0	2	4	6	10	20
cm, y	0	5				50

c) Use your values to draw a conversion graph for inches and centimetres.

d) Use your graph to find:

    (i) the number of centimetres equivalent to 16 inches

    (ii) the number of inches equivalent to 30 centimetres.

a) $y = 2.5x$		The number of cm, y, is 2.5 times the number of inches, x.	inches → $\boxed{\times 2.5}$ → centimetres
b)		4 inches = 4 × 2.5 = 10 cm	1 inch = 2.5 cm

Inches, x	0	2	4	6	10	20
cm, y	0	5	10	15	25	50

6 inches = 6 × 2.5 = 15 cm

10 inches = 10 × 2.5 = 25 cm

c) 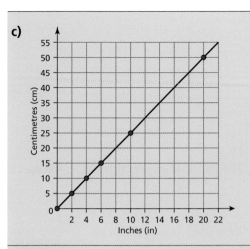	Plot the points from your table: (0, 0) (2, 5) (4, 10) (6, 15) (10, 25) (20, 50) Join the points with a straight line.	**Tip** Conversion graphs are always straight lines that go through the origin (0, 0).
d) (i) 40 cm	To convert from inches to centimetres, find the amount in inches on the horizontal axis, go up to the graph, across to the vertical axis and read off the value.	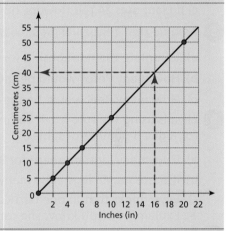
d) (ii) 12 inches	To convert from centimetres, find the amount in centimetres on the vertical axis, go across to the graph, down to the horizontal axis and read off the value.	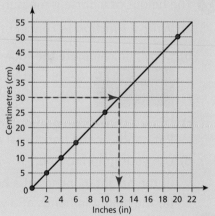

Exercise 1

1 Write the rule for each of these relationships.

    a) y is 2 more than x **b)** y is 3 less than x

    c) d is e multiplied by 5   **d)** m is n multiplied by 4

2 Write a sentence to describe each of these rules.

    a) $y = 7x$ **b)** $s = 10t$ **c)** $y = x + 6$ **d)** $v = w - 5$

3 Match the rules and their descriptions.

y is equal to 1 less than x $y = 3x$

y is equal to 5 more than x $y = x - 5$

y is equal to x multiplied by 3 $y = x + 3$

y is equal to x multiplied by 5 $y = x - 1$

y is equal to 5 less than x $y = 5x$

y is equal to x add 3 $y = x + 5$

4 Liam decorates cakes with red flowers and yellow flowers. The number of red flowers, r, he uses is 5 more than the number of yellow flowers, y.

a) Complete the relationship $r =$ _____

b) If Liam uses 10 yellow flowers, how many red flowers will he use?

c) If Liam uses 24 red flowers, how many yellow flowers will he use?

5 Assume that 1 Brazilian Real, x, is the same as 14 Argentine Pesos, y.

a) Complete the relationship $y =$ _____

b) How many Pesos would be the same as 8 Real?

c) How many Real would be the same as 700 Pesos?

6 Eliya is using the rule $L = 4t$ to find how many table legs, L, there are for a given number of tables, t.

a) How many legs are there for 8 tables?

b) If there are 48 legs, how many tables are there?

c) Eliya thinks that in one particular room there are 54 table legs. Explain how you know she can't be correct.

7 The height of a horse is measured in hands. One hand is equal to four inches.

a) Write a function to connect the number of inches, n, and the number of hands, h.

b) Copy and complete the table to convert hands to inches.

Hands, h	0	5	10	15	20	25
Inches, n						

c) Draw a conversion graph to convert hands and inches. Put hands on the horizontal axis and inches on the vertical axis.

d) Use your graph to find:

(i) the height in inches of a horse who is 18 hands tall

(ii) the height in hands of a young horse who is 42 inches tall.

8 Marcus is printing some large documents. The printer prints pages at a rate of 6 pages every minute.

a) Write a function to find the number of pages, p, printed in t minutes.

b) Copy and complete the table to show the number of pages printed.

Time in minutes, t	0	5	10	20	30
Number of pages, p					

c) Draw a graph to show the number of pages printed. Put time on the horizontal axis and the number of pages on the vertical axis.

d) The first document Marcus prints takes 16 minutes to print. How many pages does it have?

e) The second document Marcus prints has 138 pages. How long will it take to print?

9 The graph shows the currency conversion between US Dollars and South African Rand.

a) Use the graph to find:

   (i) The number of Rand equivalent to $4.

   (ii) The number of Dollars equivalent to 84 Rand.

b) Paige has $5 and Jenni has 75 Rand. Who has more money? Explain how you know.

c) Pete has $300 dollars. Explain how he could work out how much this is worth in Rand.

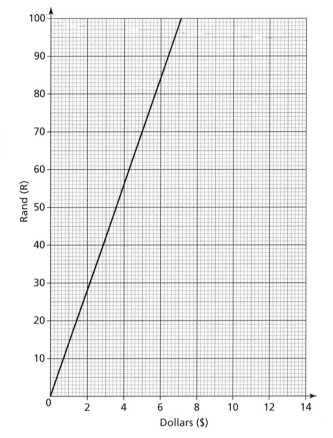

10 The graph shows the conversion between two measures of mass, kilograms (kg) and pounds (lb).

a) Use the graph to find:

   (i) the number of pounds equivalent to 10 kilograms

   (ii) the number of kilograms equivalent to 15 pounds.

b) Two newborn babies, Amir and Yazan, are weighed. Amir weighs 4 kg. Yazan weighs 7.5 pounds. Which baby weighs more? Explain how you know.

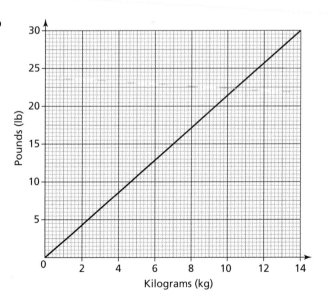

Thinking and working mathematically activity

Look up some currency conversions on the internet between your country and other countries.

Choose one conversion rate and write an approximate rule.

Use your rule to create a conversion graph.

Write four questions based on your conversion graph. Ask a partner to answer your questions.

27.2 Graphs of linear functions

Worked example 2

Here is the equation of a linear function: $y = x + 3$

a) Complete the table of values for this function:

x	0	1	2	3	4	5
y						

b) From your table, write the coordinate pairs of points that are on the line $y = x + 3$.

c) Plot the points and draw the graph of $y = x + 3$ on a coordinate grid for values of x going from 0 to 5 inclusive.

a)

x	0	1	2	3	4	5
y	3	4	5	6	7	8

The rule is $y = x + 3$, so look at each x value in turn and add 3 to get the y value.

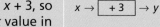

b) (0, 3), (1, 4), (2, 5), (3, 6), (4, 7), (5, 8)

A coordinate pair is (x, y), so you match up each x value with its corresponding y value.

c)

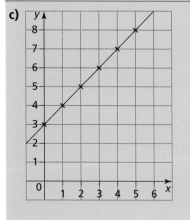

Plot each coordinate pair from your table.

Join your points with a straight line.

Tip

You need only two points to make a straight line, but you calculate at least three just to check that you have not made a mistake.

Thinking and working mathematically activity

- Use graphing software to draw the following straight lines:

  $x = 5$ $x = -2$ $x = 7$

  What do all these lines have in common?

  Describe what the line $x = 2$ will look like. Explain why this line has the equation $x = 2$.

- Use graphing software to draw a horizontal line that passes through the point (0, 4).

- What equation would give a horizontal line that passes through the point (0, −3)?

- What do the equations of all horizontal lines have in common?

Worked example 3

Identify the equations of lines A and B shown in the diagram.

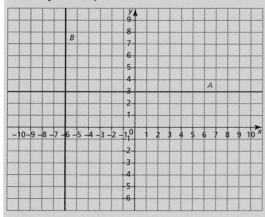

A: Pick any three points on line A.

Here, these are (−3, 3), (1, 3) and (4, 3)

Every point on A has a y-coordinate of 3.

This means the equation of line A is $y = 3$.

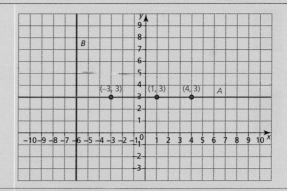

B: Pick any three points on line B.

Here, these are (−6, 9), (−6, 2) and (−6, −3)

Every point on B has an x-coordinate of −6.

This means the equation of line B is $x = -6$.

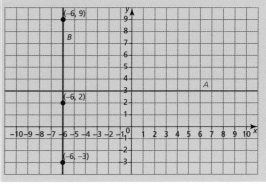

Exercise 2

1 Copy and complete the missing coordinates of points on the line $y = 4x$.

a) $(2, \_\_\_)$ b) $(5, \_\_\_)$ c) $(0, \_\_\_)$ d) $(1, \_\_\_)$ e) $(-2, \_\_\_)$

2 Copy and complete this table of coordinates for the line $y = x + 7$

x	0	1	2	3	4	5
y						

Use your table to draw the graph of $y = x + 7$.

3 Copy and complete this table of coordinates for the line $y = 2x$.

x	0	1	2	3	4	5
y						

Use your table to draw the graph of $y = 2x$.

4 a) Copy and complete this table of coordinates for the line $y = 3x$.

x	-2	-1	0	1	2	3
y						

b) Draw an x-axis from -2 to 3 and a y-axis from -6 to 9. Use your table to help you draw the line $y = 3x$.

> **Discuss**
>
> Do you notice a pattern in the y-coordinates? Can you see a link between the pattern and the equation of the line?

5 a) Copy and complete this table of coordinates for the line $y = x - 3$.

b) Draw an x-axis from -2 to 3 and a y-axis from -5 to 0. Use your table of coordinates to help you draw the line $y = x - 3$.

x	-2	-1	0	1	2	3
y						

6 a) Copy and complete this table of coordinates for the line $y = x + 5$.

x	-2	-1	0	1	2	3
y						

b) Draw an x-axis from -2 to 3 and a y-axis from 3 to 8. Use your table to help you draw the line $y = x + 5$.

7 Linda is drawing the graph $y = x + 4$. She says the point $(4, 8)$ is on her line.

Kristi is drawing the graph $y = 2x$. She also says the point $(4, 8)$ is one her line.

Can they both be correct?

Draw both graphs on the same pair of axes and use your diagram to help you explain your answer.

8 Give the coordinates of any three points on each of these lines.

a) $y = 2$ b) $x = 5$ c) $y = 0$ d) $y = -4$

9 **a)** Give the coordinates of three points on each line A to H in the diagram.

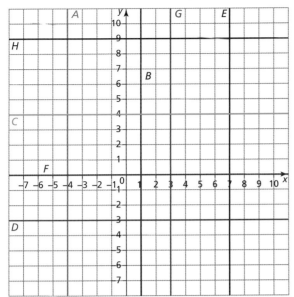

b) Use your answers to part **a)** to give the equation of each line.

10 On the same set of axes, draw the graphs of:

a) $y = -4$ **b)** $x = -2$ **c)** $x = 0$ **d)** $y = 3$

11 **a)** Which horizontal and vertical lines does the point (3, 1) lie on?

b) Which horizontal and vertical lines does the point (a, b) lie on?

12 Write the following points in the correct position in the table.

$(-1, 7)$ $(2, -7)$ $(5, 7)$ $(5, 2)$ $(7, 5)$

	On $x = 5$	Not on $x = 5$
On $y = 7$		
Not on $y = 7$		

13 Write the equations of three vertical and three horizontal lines in their correct places on the Venn diagram.

Make sure you include at least one equation for each part of the Venn diagram.

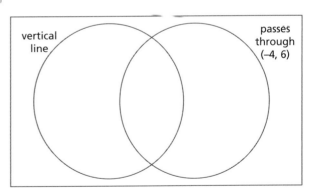

14 Simone says that all the lines of the form $y = x + c$ pass through (0, 0), the origin.

Max says that the only line passing through the point (2, 8) is the line $y = 4x$.

The statements made by Simone and Max are not correct.

Change their statements to make them true.

15 **Technology question** Use graphing software to plot the graphs of:

a) $y = x + 5$ b) $y = x + 8$ c) $y = x - 3$

What do you notice about where each of your graphs crosses the y-axis?

27.3 Graphs in real-life contexts

Key terms

A graph which shows a journey in terms of distance and time is called a **travel graph**.

Graphs can also be used to show the **rate of change** of something, such as temperature over the course of a day. This type of graph can be used to help show or describe what is happening over a period of time.

Worked example 4

Annie goes to the library to study. She cycles to the library from home, spends some time doing her homework at the library and then cycles home.

The graph represents her journey to the library, her time spent at the library and her journey home.

a) When did she leave home?

b) How far from her home is the library?

c) How long did she spend at the library?

d) How long did it take her to cycle home?

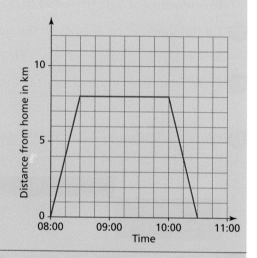

a) Annie left home at 08:00 as this is when the graph line moves away from the horizontal axis.

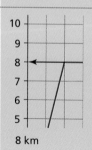

b) At 08:30 the graph line becomes horizontal, showing that Annie's distance from home is not changing, so she has arrived at the library. This shows that the library is 8 km from her home.

c) The graph line is horizontal between 08:30 and 10:00, which is 1 hour and 30 minutes spent in the library.

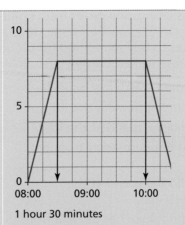

1 hour 30 minutes

d) Annie leaves the library at 10:00. You can see when she arrives home when her distance from home is 0. This is where the graph line meets the horizontal axis. This is at 10:30, so Annie cycled home in 30 minutes.

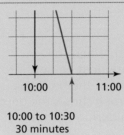

10:00 to 10:30
30 minutes

Tip

When reading a graph, always look at the scales first and work out what each square represents. You can then check it by counting on. For example, if you look at this section of a graph, there are 5 squares between 10 and 20. The difference between 10 and 20 is 10. So, 10 divided by 5 is 2 which means each square represents two units. Count on from 10 to 20 in 2s for each square to check. The arrow points at 16.

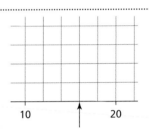

Exercise 3

1 The graph represents Paolo's journey as he cycles from home to the shop and back.

State whether the following statements are true or false. If they are false then give the correct answer.

a) Paolo leaves home at 13:00.

b) Paolo spends 45 minutes at the shop.

c) The shop is 7 km from his house.

d) It takes him 1 hour to cycle to the shop and back.

e) Copy and complete these sentences.

Paolo leaves home at _____ and arrives at the shop at _____ .

He leaves the shop at _____ .

He leaves home at _____ and arrives back home at _____ so he is away from home for a total of _____ hours.

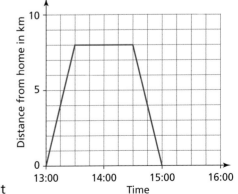

2 Carey is going skating. She cycles from home to the ice rink. On the way home she stops at the supermarket to buy some food for dinner. The graph represents her journey.

a) What time did Carey leave home?

b) How far from her home is the ice rink?

c) How long did she spend at the ice rink?

d) How far is the supermarket from her home?

e) How long did she spend at the supermarket?

f) How long did it take her to cycle from the supermarket to her home?

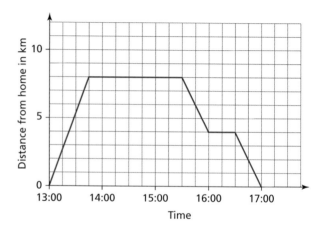

3 Alvarez is filling the bath. When the bathtub has enough water in it he gets into the bath. He washes, gets out of the bath and empties the bath. The graph shows the depth of the water in the bath during this time.

a) What is the depth of water in the bath just before Alvarez gets in?

b) What time does Alvarez get into his bath? How can you tell this from the graph?

c) What is the depth of the water when Alvarez is in the bath?

d) How long does Alvarez spend in the bath?

e) How long does it take for the bath to empty?

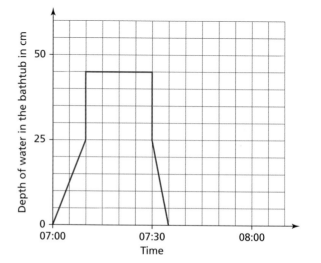

4 The graph shows the temperatures in London on a day in June.

a) What was the temperature at 03:00?

b) What was the rise in temperature between 06:00 and 09:00?

c) What was the temperature at 12:00?

d) How much did the temperature rise between 12:00 and 15:00?

e) Did the temperature rise at a greater rate between 03:00 and 06:00 or between 06:00 and 09:00? How can you tell?

f) Describe what happened to the temperature after 15:00.

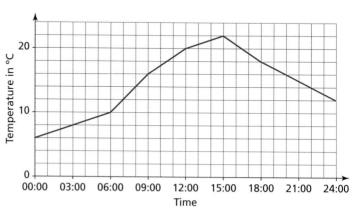

5 Adele leaves home at 08:00 to walk to school which is 2 km away. She takes 15 minutes to walk the first kilometre. She then stops to talk to her friend Jenna for 10 minutes. She then walks the rest of the way to school in another 15 minutes. Show her journey on a copy of the graph to the right.

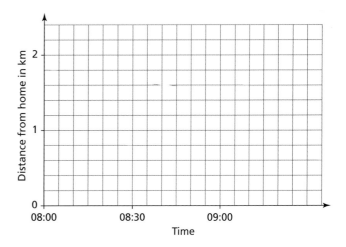

6 A plane flies from London to Paris. It stops in Paris for 30 minutes before flying on to Rome. This part of the journey is shown on the travel graph.

The plane then stops for 36 minutes in Rome before making a $\frac{1}{2}$ hour return flight to London.

a) Copy and complete the travel graph to show the entire journey.

b) What time does the plane arrive back in London?

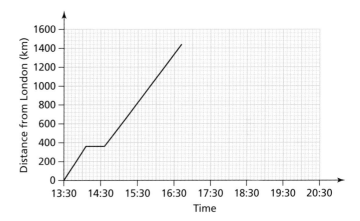

7 A bus sets off from Town A to Town B. It makes two stops along the way.

The journey is shown on the travel graph.

a) For how long does the bus wait at the first stop?

b) What is the distance between the first and second stops?

c) What time does the bus arrive at town B?

d) When is the bus travelling the fastest? Explain how you know.

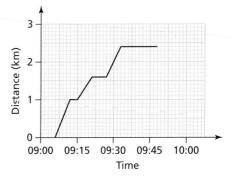

8 Three containers P, Q and R are filled up from the top with a steady flow of water.

The graphs show the depth of water in each container over time.

Match each graph to the container it represents. Explain how you know.

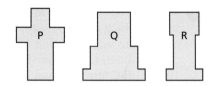

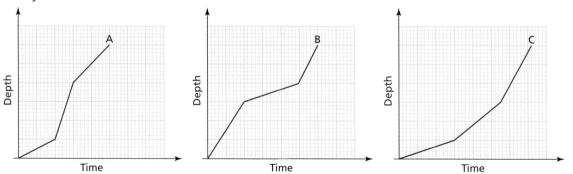

> **Thinking and working mathematically activity**
>
> Create a travel graph for a journey of your own choice. For example, you could draw a journey to school or a trip to the shops.
>
> Swap your travel graph with a partner. Ask your partner to write a description of the journey based on the shape of your graph.

Consolidation exercise

1 Which of these is the odd one out? Give a reason for your answer.

A:	B:	C:	D:
the x-axis	$y = 4$	the y-axis	$y = -7$

> **Tip**
>
> Draw the graphs each one on the same axes.

2 Which of these points lie on the line $y = 8x$?

a) (4, 32) **b)** (12, 100) **c)** (27, 216) **d)** (48, 390) **e)** (0, 0)

3 Draw a pair of axes going from −6 to 6.

Draw these lines on your axes:

a) $y = -2$ **b)** $y = x - 3$ **c)** $y = 4$ **d)** $y = x + 4$

e) What is the name of the shape enclosed by your lines?

4 Copy and complete the table by putting the following coordinate pairs in the right places.

(4, 5)   (−2, 5)   (4, 8)   (0, 8)

	On $x = 4$	Not on $x = 4$
On $y = 5$		
Not on $y = 5$		

5 One Australian Dollar is equivalent to 5 Chinese Yuan.

  a) Copy and complete the table.

Australian Dollars	0	5	10	20	30	50
Chinese Yuan						

  b) Draw a graph to show the conversion between Australian Dollars and Chinese Yuan.

  c) Use your graph to convert 40 Australian Dollars to Chinese Yuan.

  d) Emma has 24 Australian Dollars and Tia has 125 Chinese Yuan.
  Whose money is worth more? Explain how you know.

6 The graph shows the conversion between miles and kilometres.

  Use the graph to find:

  a) The number of kilometres in 60 miles.

  b) The number of miles in 100 kilometres.

  c) Explain how you could use your graph to find the number of kilometres in 300 miles.

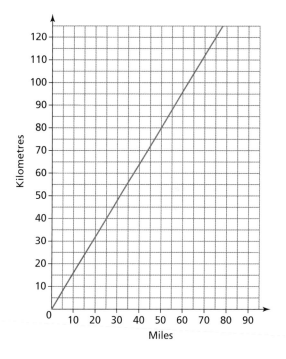

7 Petra walks to the swimming pool from home. She then swims with her friends and walks home. The graph represents her journey.

  a) What time does Petra leave home?

  b) How far from her home is the swimming pool?

  c) How long does she spend at the swimming pool?

  d) How long does it take Petra to walk home?

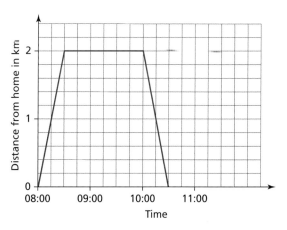

End of chapter reflection

You should know that...	You should be able to...	Such as...
Conversion graphs are straight line graphs that convert between different units of measurement.	Draw and interpret a conversion graph.	Draw a graph to convert between pints and litres. Use your graph to work out how many pints there are in 5 litres.
A linear function can be represented by a straight line graph.	Plot a graph of a given equation.	Plot a graph of $y = 2x$ for values of x between 0 and 5.
Points on a line all fit the rule for that line.	Say whether a coordinate pair lies on a given line.	Does the coordinate (2, 6) line on the line $y = 3x$?
Horizontal lines have equations of the form $y = a$ and vertical lines have equations of the form $x = a$.	Draw a horizontal or vertical line from its rule.	Draw the line $x = 3$.
Graphs can represent real-life situations.	Draw and interpret graphs representing real-life situations.	Draw a travel graph to represent your journey to school.